SCAVNICKY: PORTRAIT OF AN ANTHRACITE FAMILY
NELSON MORRIS, PHOTOGRAPHER

June 21 through July 26, 1992

Sordoni Art Gallery
Wilkes University
Wilkes-Barre, Pennsylvania

This exhibition is funded in part by a grant
from the Pennsylvania Council on the Arts.
Additional support was provided by
Pagnotti Enterprises, Inc. and Carbon Sales, Inc.

ISBN 0-942945-02-6

Fifteen hundred copies of this catalog were printed by
Design Systems Printing, Wilkes-Barre, Pennsylvania.
The typefaces used are Clearface and Clearface Black,
originally drawn by Morris Fuller Benton in 1907.

INTRODUCTION AND ACKNOWLEDGMENTS

This series of thirty eight photographs of Andrew Scavnicky, a miner in the Jeddo-Highland Mine No. 7 near Hazleton, Pennsylvania, were found not in a local archive but at the Library of Congress in Washington, D.C. Taken in 1942, the portfolio is clearly important to Northeastern Pennsylvania as a historical document not only of the anthracite industry, which played such an important role in this region's history, but also of the human element personified by the Czechoslovakian miner who was its subject. Reproduced together for the first time in this catalog, these photographs are remarkably complete in the story they tell.

Annie Bohlin discovered the photographs, and the original captions written to accompany them, in 1982 while doing research for another project. Subsequently, she interviewed Mr. and Mrs. Andrew Scavnicky, Jr., and later brought the material to my attention. She has served as guest curator for this exhibition, which is the first to provide a public viewing of the photographs. Asked to tell the human side of the story described in our title, *Scavnicky: Portrait of an Anthracite Family,* Bohlin produced one of the accompanying essays which follow. Her insightful design of this catalog has brought the project together into an important and handsome resource.

Dan Perry, Director of the Anthracite Heritage Museum in Scranton, graciously agreed to contribute an essay which would place the photographs in a historical context as they relate to the history of anthracite mining. His comments are an invaluable part of this catalog. Further, his experience in exhibitions which focus on several dimensions of interpretation, as this one does in addressing the aesthetic, historical and social aspects of these photographs, brings an important aspect of institutional collaboration to our project, for which we are grateful. Comments from Chester Kulesa, Curator of the Anthracite Heritage Museum; Professor Harold Aurand, of the Penn State Hazleton Campus; and Frank Dushanko, of Hazleton, were also helpful.

Kay Reese, former executive secretary of the American Society of Magazine Photographers provided assistance in research on the photographer, Nelson Morris, whose name came to light very late in the project. It is ironic that had the ASMP, an organization Morris helped found in 1945, and its Code of Minimum Standards been in force when the Scavnicky photographs were taken, we would have known the photographer's name from the start. The essay I have contributed here will, I hope, shed light on that aspect of our story.

Fourteen of the photographs appeared in 1942 in the July issue of *Click* magazine, a national monthly photographic magazine, to illustrate a story titled "Coal Family: Andrew Scavnicky, miner, helps fuel the factories of war."

Judith H. O'Toole
Director, Sordoni Art Gallery
Wilkes University

NELSON MORRIS AND PHOTOGRAPHERS' RIGHTS

Photographer Nelson Morris was hired in 1942 to take photographs representing a typical day in the life of an anthracite miner. These were used to illustrate a story, "Coal Family: Andrew Scavnicky, miner, helps fuel the factories of war," written by Mark S. James, then Secretary of the Pennsylvania Department of Commerce, intended to generate civilian support for the war effort by showing how ordinary citizens affect military efforts. The obvious propagandizing purpose of the article is reinforced by the photographs which show Andrew Scavnicky going about his daily routine. The stocky, mustachioed Scavnicky is an archetype of the immigrant worker dedicated to job and family. Heroically depicted among fellow miners, Scavnicky stalwartly descends into the earth to perform his backbreaking work. When shown at home, he is the kindly grandfather with grandson on his knee presiding over the family breakfast table.

Morris succeeded, however, in making photographs that are strong and interesting in composition as well as message. The aesthetic concerns are addressed equally with the primary concern of story telling. Especially compelling are the portraits of Scavnicky, his family, and co-workers.

When *Click* magazine ran the photographs for which Nelson Morris had been commissioned and paid, the editors did not credit him by name on the pages where the images appeared. In fact, when *Click* later gave permission to the United Mine Workers Journal to reproduce one of the images the credit line read "Photograph courtesy of *Click* magazine." The client owned all rights to the negative and therefore to the reuse of the image with no obligation to provide credit, or further payment, to the photographer.

At the time, this lack of proper credit to the photographer was standard practice for magazines and journals, including the larger ones – *Collier's, Esquire, Life* and *Time.* Even publications which catered to the photography profession such as *Popular Photography* and *Modern Photography,* which one might think would exhibit more sensitivity to their audience, did not have policies which called for the acknowledgment of their contributing photographers.

This was due in part to the relative youth of the medium of photography and the professions which were developing around it. It was complicated further by such complex ideas as art versus technology and the issues of ownership and authenticity raised by multiple images and their infinite potential for reproduction. An analogy in another medium is provided by the authors of the first prints, lithographs and chromoliths, made for mass reproduction at the turn of the century who were often not acknowledged in the final form. Further confusing the issue as it relates to photography, however, is the notion that the literal "hand" of the artist is not present. Especially where photographs are reproduced in the print media, it becomes the mind of the artist working through the technical instrument of the camera that is most present – a rather abstract issue which is still being debated.

However, a heightened awareness of the interpretive role of image makers, be they printmaker or photographer, has changed our perception of attribution. Our contemporary perspective sees photographers as individuals and artists rather than technicians and we are accustomed to seeing that small type running vertically up the side of published photographs, naming the photographer.

Nelson Morris, the author of the Scavnicky photographs, played a crucial role in helping modern audiences arrive at these conclusions as a founder of the American Society of Magazine Photographers. Established to protect the rights of its members and clarify the relationship between photographer and publisher, the ASMP drafted its Code of Minimum Standards, which radically changed this relationship. The Code also standardized practices, because part of the problem had been that each magazine had different policies affecting photographers.

Before the Code of Minimum Standards, photographers had often been required to advance their own expenses, travel to an assignment, submit finished pictures, and be paid only if an image was used and even then only for perhaps one photograph out of dozens taken. Photographers also had no subsequent right to the image as the negative became the property of the magazine to reuse at will. A writer could maintain copyright after a story was run in a periodical and therefore accept royalties should it be published in book form, but a photographer did not have the same right.

Adopted by the ASMP in 1952, the Code laid out in fifteen clauses rules regarding compensation, specific use of images, ownership of negatives and, importantly, the use of credit lines acknowledging the photographer by name.

Nelson Morris's name did not surface until well into the research for this exhibition and catalog. It does not appear anywhere on the documentation accompanying the original photographs which were given to the Library of Congress by the United States Office of War Information. Those who knew of the photographs guessed that they were made by someone of the stature of Margaret Bourke-White, thereby attesting to their quality. It is in recognition of Nelson Morris's efforts on behalf of all photographers that this portfolio of images is fully reproduced here with proper credit at last accorded the artist.

Judith H. O'Toole
June 1992

HARD COAL HERITAGE

Anthracite coal was king. For nearly a century it reigned supreme over a ten county realm in Northeastern Pennsylvania. So dominant was the single industry that today this area is still commonly referred to as the Anthracite Region.

The search for black diamonds stimulated widespread immigration into the Region and ethnic communities were created overnight. Hundreds of thousands entered the earth in pursuit of it, and tens of thousands lost their lives in the process. Coal generated vast fortunes for some, but the majority of those associated with the industry found only a modest income in exchange for untold toil in a dangerous and alien environment.

The first recorded use of "stone" coal came in 1769. It was probably obtained from readily accessible outcrops lying on the surface. Initially, anthracite found only limited, and local, use as a fuel source. Fifty years passed before demand outpaced supply and an industry devoted solely to the extraction of this mineral was born.

Anthracite fueled America's Industrial Revolution. During the early years, outside markets were limited to New York and Philadelphia. Canals and railroads were constructed to link the coal fields of Pennsylvania's mountainous northeast to these metropolitan areas. By mid 19th century, as transportation systems improved and further uses for anthracite developed, demand for hard coal increased dramatically and large scale mining was begun.

In the case of the individual miner, techniques employed by Andrew Scavnicky in 1942 differed little from those used during the previous century. Although by World War II numerous advances had greatly improved both safety underground and processing above, the industry still relied almost entirely on the knowledge and strength of the lone miner and his laborer using basic hand tools.

Anthracite mines were dangerous places with injuries and deaths common occurrences. Between 1869, when the Commonwealth of Pennsylvania began keeping records, and 1900, some 10,000 men and boys lost their lives while working in the coal fields. The majority of these fatalities came, not as a result of widely publicized disasters, but from countless small rock falls or machinery accidents which claimed one or two men at a time.

The anthracite industry reached its peak in 1917. During that year, some 100 million tons were extracted and processed by a workforce totaling nearly 160,000. The majority of these employees lived in "patch" towns which sat huddled in the shadow of a looming coal breaker. The lives of both mine workers and their families were intimately tied to the coal company. Often, they lived in a company owned house and were compelled to purchase supplies at the company store "on account." Rent and other expenses owed to the coal company were deducted from the miner's pay. The colliery whistle regulated daily existence for residents of the patch. It summoned men to work, served as a community's first notice of disaster underground, and provided a means of telling time in an occupation where the sun never shined.

Though coal mining proved to be a hazardous occupation, it also represented opportunity. Hundreds of thousands came to Northeastern Pennsylvania seeking employment specifically in the mines and breakers. Prior to the industry's post World War I decline jobs were plentiful for those with strong backs and an ability to endure long hours amidst choking smoke in a wet, cold and cramped space. Coal companies provided a place to work and the means for getting the coal to the surface, but the miner's gross wages were based on what he and his laborer were able to blast free and load onto waiting coal cars. In this respect, they enjoyed a degree of independence over their brethren engaged in other above ground occupations. Generally, the laborer was paid by the miner. In 1942, when Scavnicky and his family were captured on film, mine workers extracted an average of nearly three tons of coal per man per day.

World War II provided a momentary reprieve from the steady decline which had gripped the industry during the preceding two decades. Demands for anthracite brought on by the war effort were temporary and evaporated soon after the conflict ended. By mid-century, anthracite had lost the majority of its market share to other energy sources, and although abundant deposits remained, they became increasingly expensive to mine.

The post-war period witnessed the closing of mine after mine. Jeddo #7, where Scavnicky worked, survived the widespread decline due to the feasibility of strip mining in the immediate vicinity. Due primarily to its cost effective nature, strip mining became the dominant means of extracting hard coal during the second half of the 20th century.

The legacy of anthracite has been one of devastation. Environmental degradation and widespread unemployment are but two of the realities which followed in the wake of the industry's demise. It is only recently that many of the problems related to mining are being addressed. The most notable reflection of the coal era lies within the people themselves. Those who lived through the days of king coal possess a tough and resilient character born of a hard life in the pursuit of hard coal.

Daniel K. Perry
June 1992

GIANTS IN THE EARTH

In 1905 Andrew Scavnicky and his wife came to Hazleton, Pennsylvania, from Czechoslovakia. Brothers of Mrs. Scavnicky had already settled in Hazleton, and they sponsored the immigrant Scavnickys. After first working on the foundation for St. Joseph's Church in his new community, Andrew found work at the Jeddo-Highland Mines, the Markle family's workings, and he and his wife moved into one of the company houses at Ebervale #2 and started a family of ten children, six boys and four girls. Andrew worked underground as a miner until after World War II, when stripmining began to replace underground mining, and he continued to work with the same company, in surface jobs, until his death in 1952 at the age of 77.

From 1925 to 1973 the Scavnickys' eldest son, Andrew, also worked in the Markle mines, starting with two years as a "mucker," then being a miner's assistant, miner, shiftleader, fire-boss, safety inspector, and mechanical-mining foreman underground before moving to "outside" work in the Jeddo #7 breaker in 1961.

Early in 1942 our war effort was roaring into high gear. Nelson Morris came to Hazleton and made these photographs of a Slovak anthracite miner and his family for the Overseas Branch Exhibit Section of the U. S. Office of War Information. Were they intended only to illustrate and personalize a propaganda article on the deployment alternatives for our principal energy resources (*Click*, vol. 5, no. 7, July, 1942)? Or were they used in other ways here or abroad? So far this isn't clear; but in 1944 Exhibit #23, titled "Anthracite Coal Miner" - thirty eight photographs with captions and a short introduction - was transferred to the Library of Congress and became part of the nation's public records.

Early in 1982 I spent a day in the Division of Prints and Photographs of the Library of Congress looking for additional illustrations for the second edition of our historical society's ethnic cookbook (*The Whole Valley Cookbook*, Wyoming Historical & Geological Society, 1982). Toward the end of the afternoon I opened a cardboard box marked "Lot 9442, Anthracite Coal Miner" and found Andrew Scavnicky and his family.

There were thirty eight photographs rich in personal and technical detail. They were beautiful photographs. Scenes above- and underground showed the anthracite habitat, the family and its environment and activities, and the mining team at various stages of its workday. There were clear descriptive pictures in the mining chamber as the men drill, blast, shovel, and haul the coal out. There were prints, but no negatives. The history of this graphic essay was mysterious. The photographer was not identified. The date given was 1942, and the source was the Office of War Information. Transfer to the Library had been made in 1944. I wondered how many people had opened this box since then; none of the images looked familiar to me. Their visual and narrative strength was a shock. The Library had no additional information, and a few inquiries to the National Archives and other institutions led nowhere. 1942 seemed a long time ago to me then. My notes and some photocopies went into a folder, and one of the photographs (#6) was used with a new Slovak recipe for the cookbook.

In 1983 Grace Mumaw Snyder, who grew up in and near Jeddo, helped me contact the Scavnicky family. That October Grace accompanied me to visit Mr. and Mrs. Andrew Scavnicky, who have lived at Middletown, between Ebervale and Jeddo, since 1941. By this time I had requested a complete set of the pictures from the Library of Congress and had an album of prints to take along. We spent the morning with the Scavnickys around the table in their pleasant kitchen. There I first saw the article "Coal Family" in *Click*, and the Scavnickys first saw the other pictures in the series. The magazine provided, on the contents page, the name of the photographer: Nelson Morris.

That morning there was talk of mines and mining as it was done in our region, of obligations and dangers, of inmigration and outmigration, of miners and their families. The past fifty years have seen great changes in the techniques and economics of anthracite mining. No longer do men battle the elements underground. Now the giants are machines. And today the Jeddo #7 breaker is fed with anthracite torn from the surface in eighty-ton bites.

Annie Bohlin
June 1992

(The following introduction and captions are part of Lot 9442 in the Division of Prints and Photographs of the Library of Congress.)

ANTHRACITE COAL MINER

One of the most valuable American resources is coal, the mineral used in the manufacture of graphite, batteries and explosives, used to produce gas for sheet mills, steel treating furnaces, and chemical industries, used to heat American homes and drive American locomotives. Coal, one of the sources of power without which the nation's factories would be idle, has, moreover, a great potential value as an effective substitute for coke and petroleum.

The continental United States contains 51% of the world's coal supply, scattered under one fourth of the land. It is mined in 25 states but Pennsylvania, West Virginia, Illinois, and Kentucky produce four fifths of the total national output.

Buried beneath 484 square miles of Pennsylvania mountains and meadows lie four canoe-shaped beds of anthracite coal - 15,000,000,000 tons. This is the story of Andrew Scavnicky, one of the men whose job it is to dig the dusty mineral from the subterranean rock and bring it to the surface where its potential power can be realized. Andrew Scavnicky, a Czech by birth, has spent 36 years digging in the mines of Pennsylvania, and he and his wife, whom he married before coming to this country, have raised ten children on American soil. Although 67 years old, Andrew is working vigorously today to produce the fuel of war.

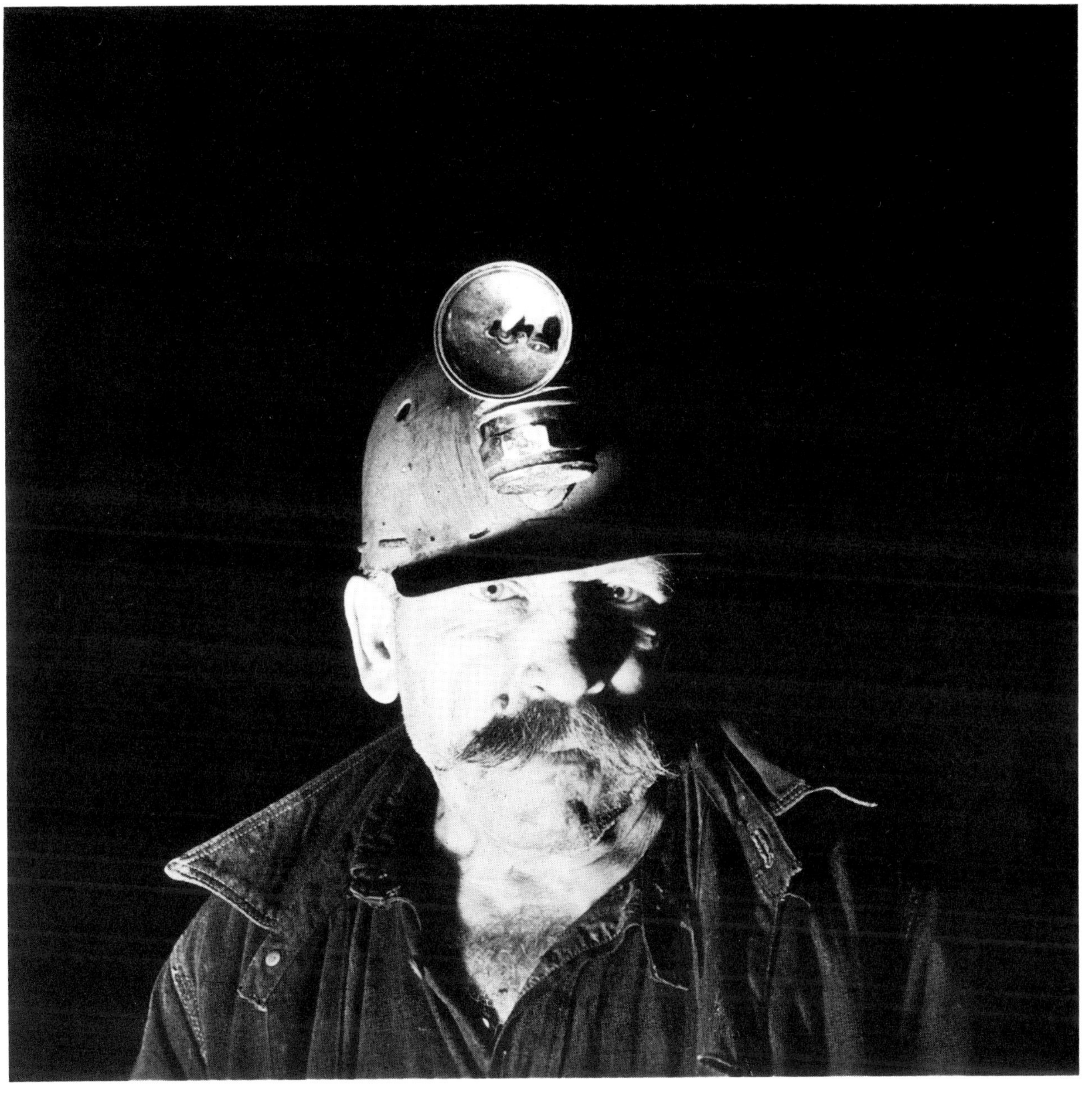

1. Andrew Scavnicky, American miner of Czech birth, is one of the army of strong men to burrow in the darkness of U.S. anthracite coal mines to assure democracy's future. Scavnicky was 67 when these pictures were taken in 1942 but still hale and hearty.

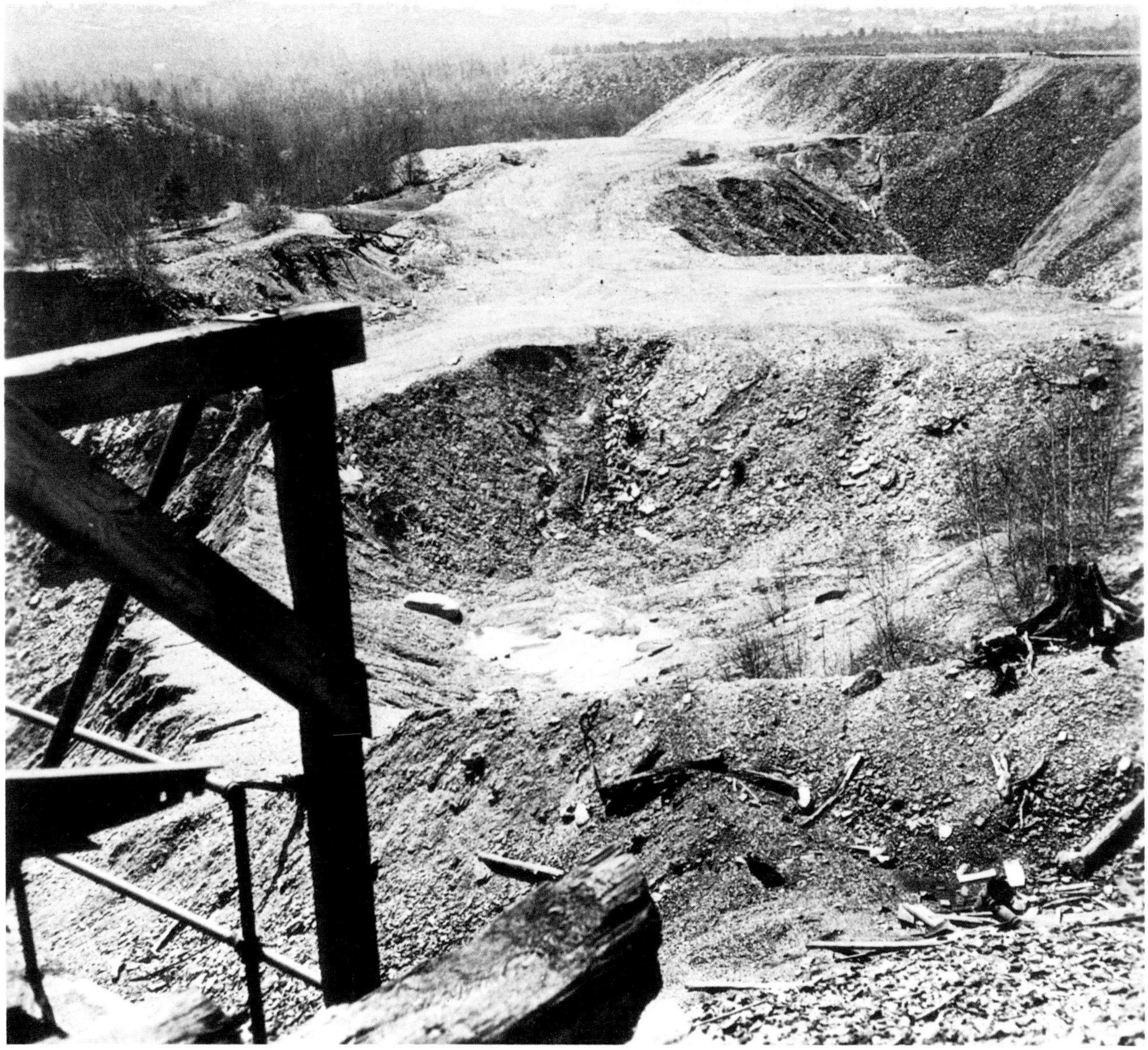

2. Growing mountains of slag are piling up around the pit-head. Such industrial landscape is strictly limited in extent. Over the next hill green and pleasant fields and woods stretch out.

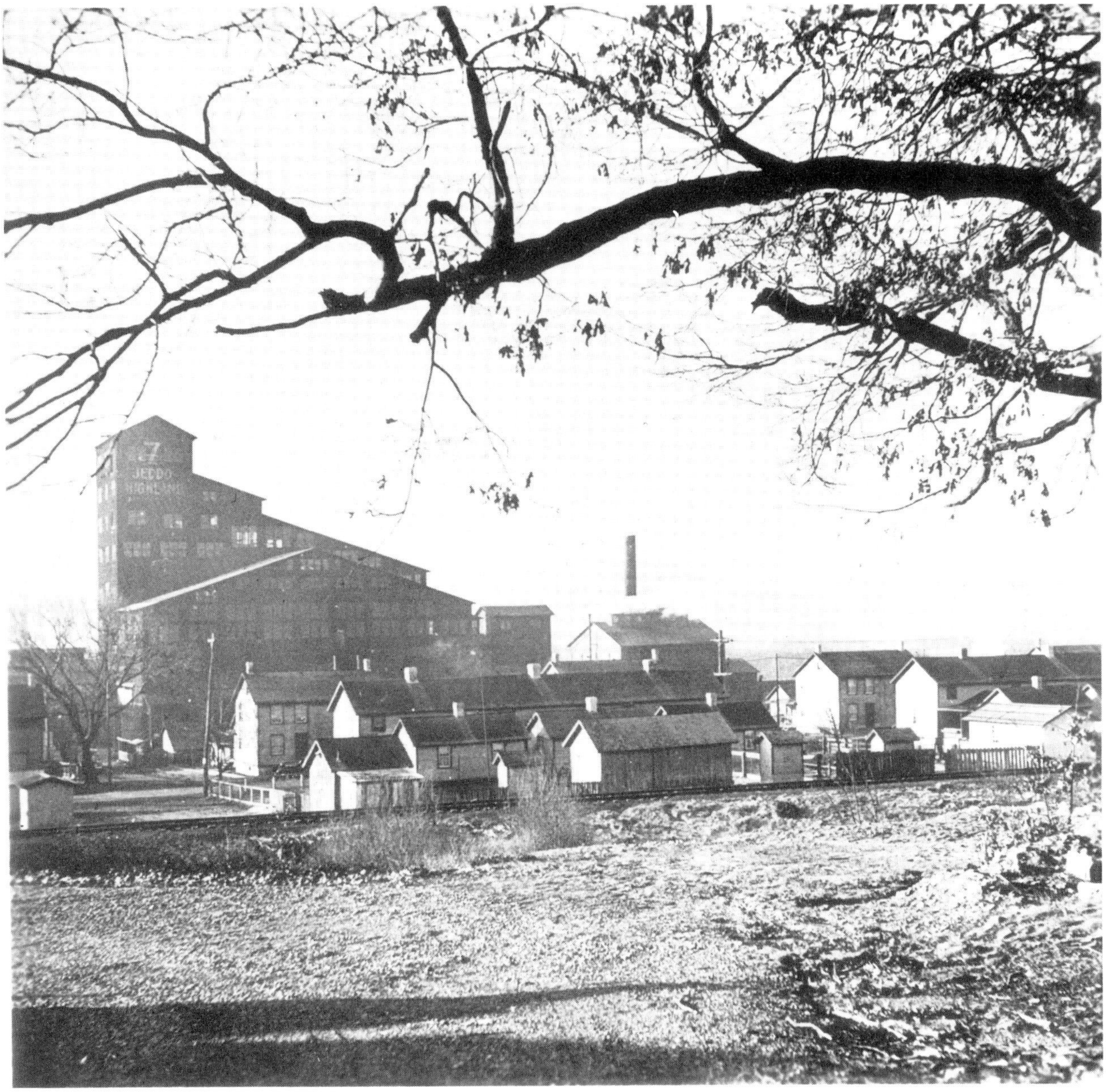

3. Pit head buildings, Jeddo-Highland Mine No. 7, Hazleton, Pennsylvania. In the tall building the breaker coal is screened (sorted by size), washed and loaded for shipment by railway. A community of company houses clusters nearby. Andrew Scavnicky and his family live in one of them.

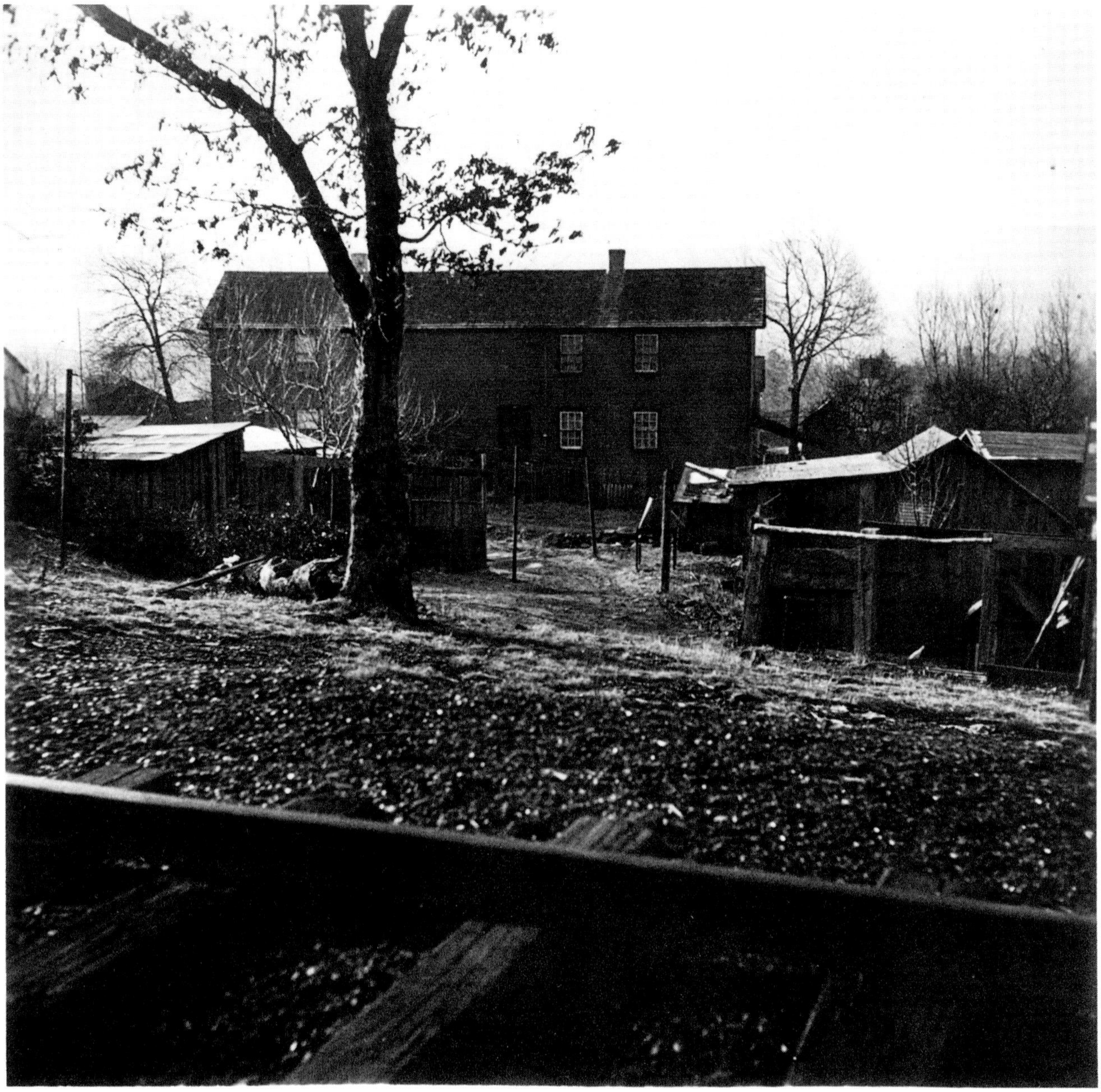

4. The Scavnickys have lived in this house since 1906. Though depressingly gray outside, it is snug and comfortable within.

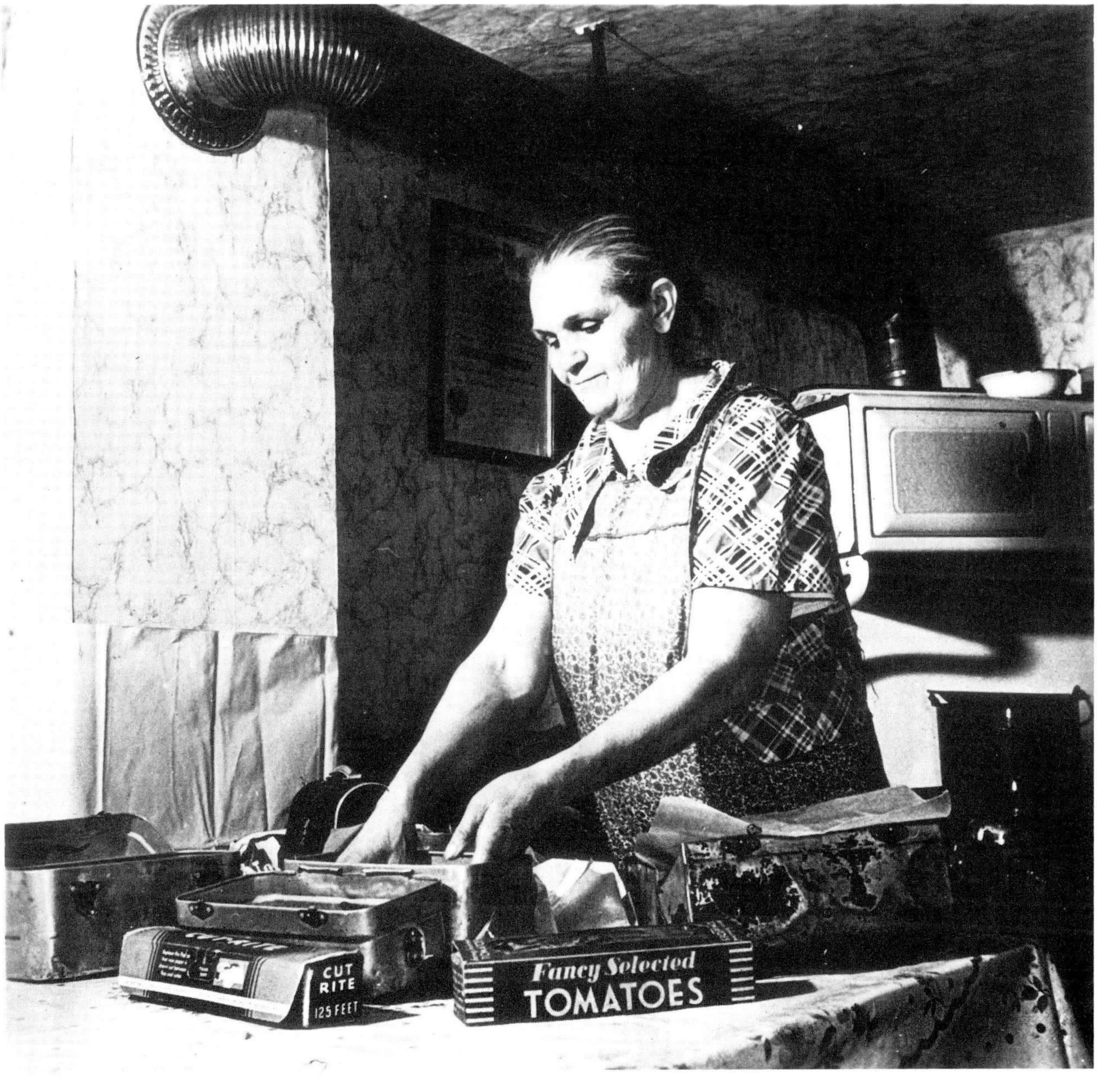

5. Mrs. Scavnicky packs her husband's lunch. She and Andrew were married in the old country.

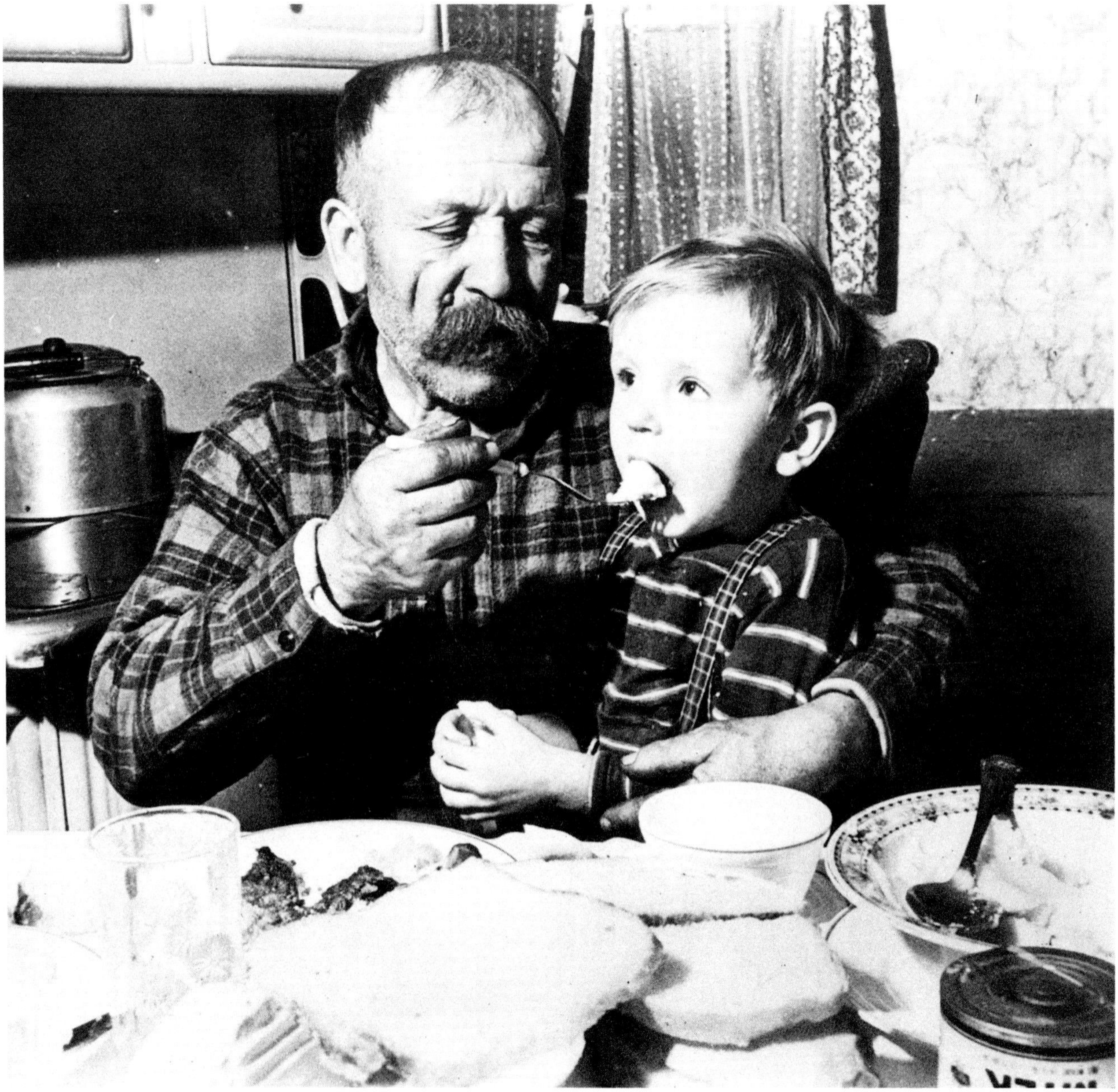

6. While waiting for his lunch box Andrew helps young Jack with his breakfast. Jack is one of ten grandchildren. Andrew and his wife have ten children, are a devoted family couple.

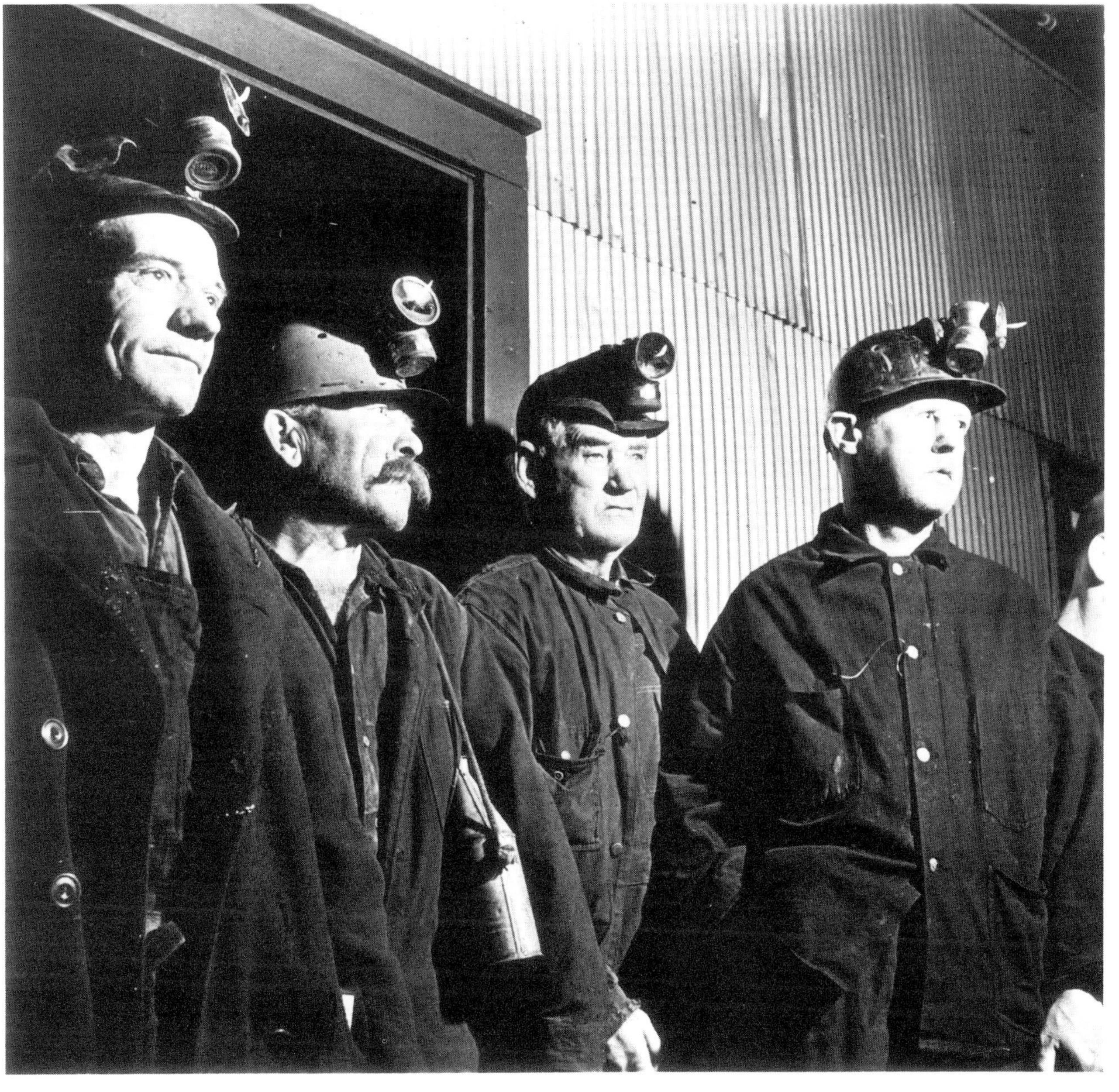

7. With fellow miners in their working clothes, carbide lamps attached to their helmets, Andrew Scavnicky waits at the pit head for the man car.

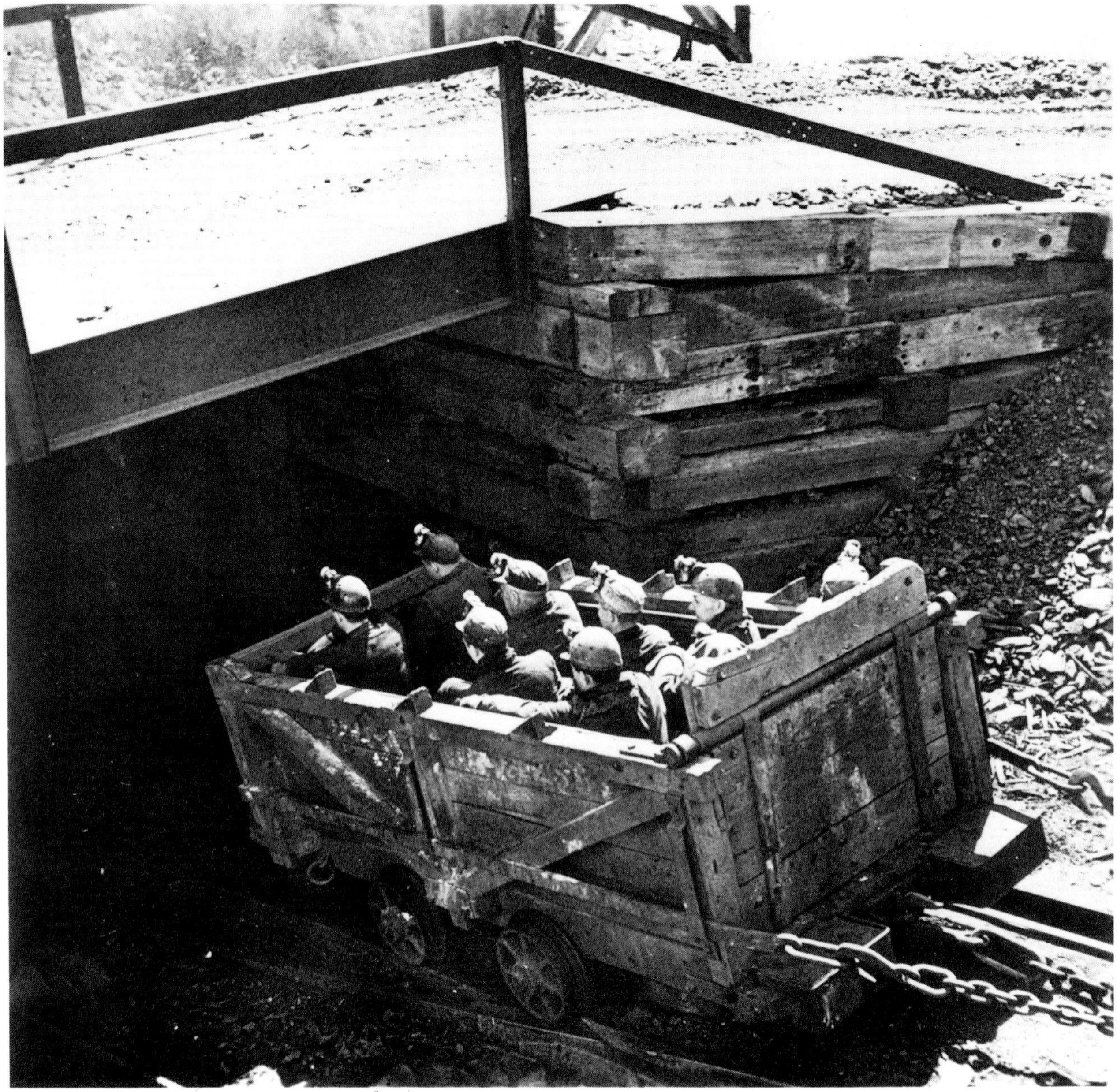

8. Away they go. The man car takes 9 miners to their subterranean labor. Andrew has lived most of his life underground.

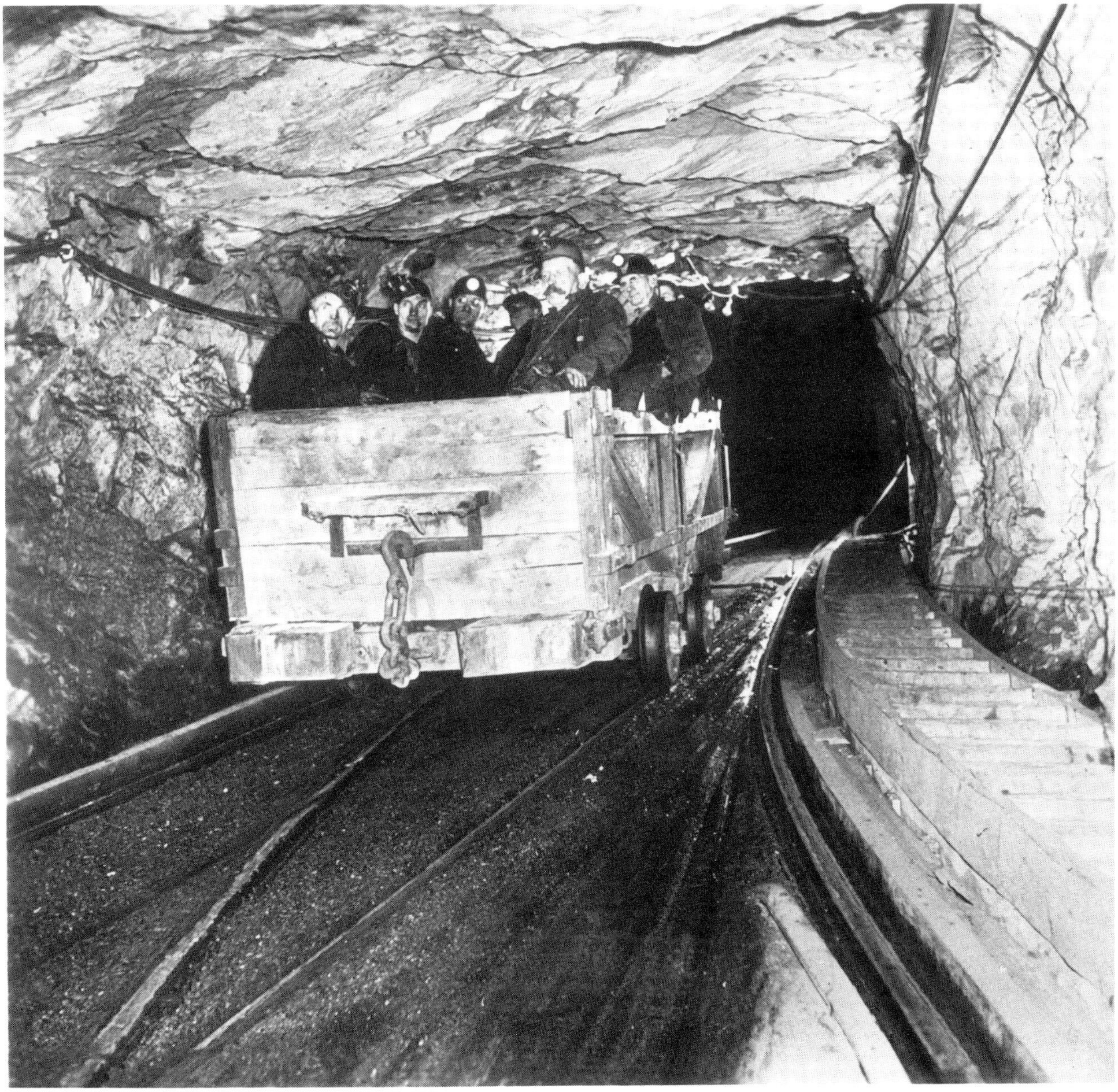

9. Still riding to work at the company's expense. Through tunnels like this air and electricity, as well as men and coal, are carried.

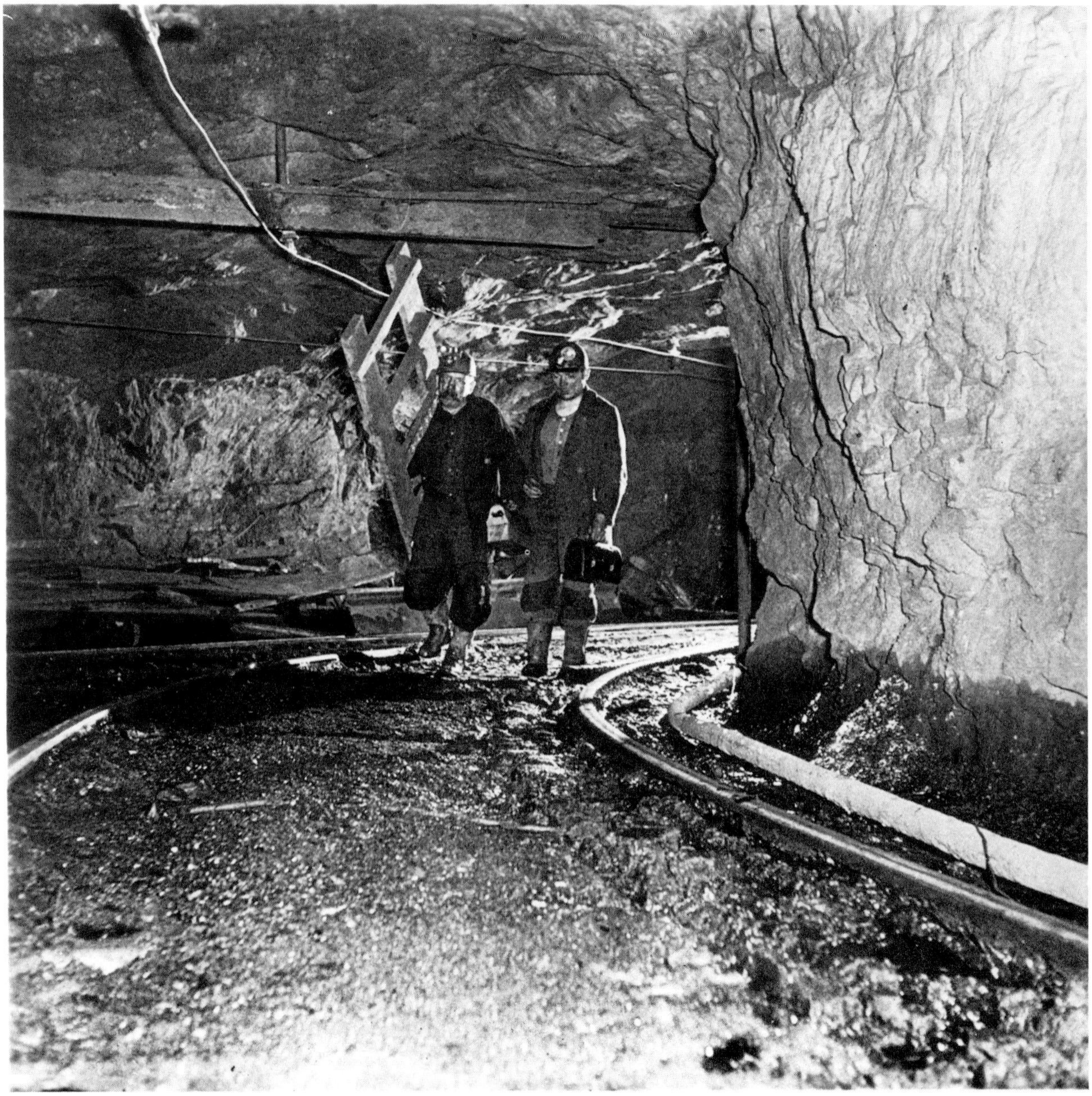

10. Andrew Scavnicky and his helper John Mikula are leaving the main chamber and walking toward the new vein they are digging. The ladder-like contraption can be quickly placed to brace sagging timbers if operations tend to start a cave-in.

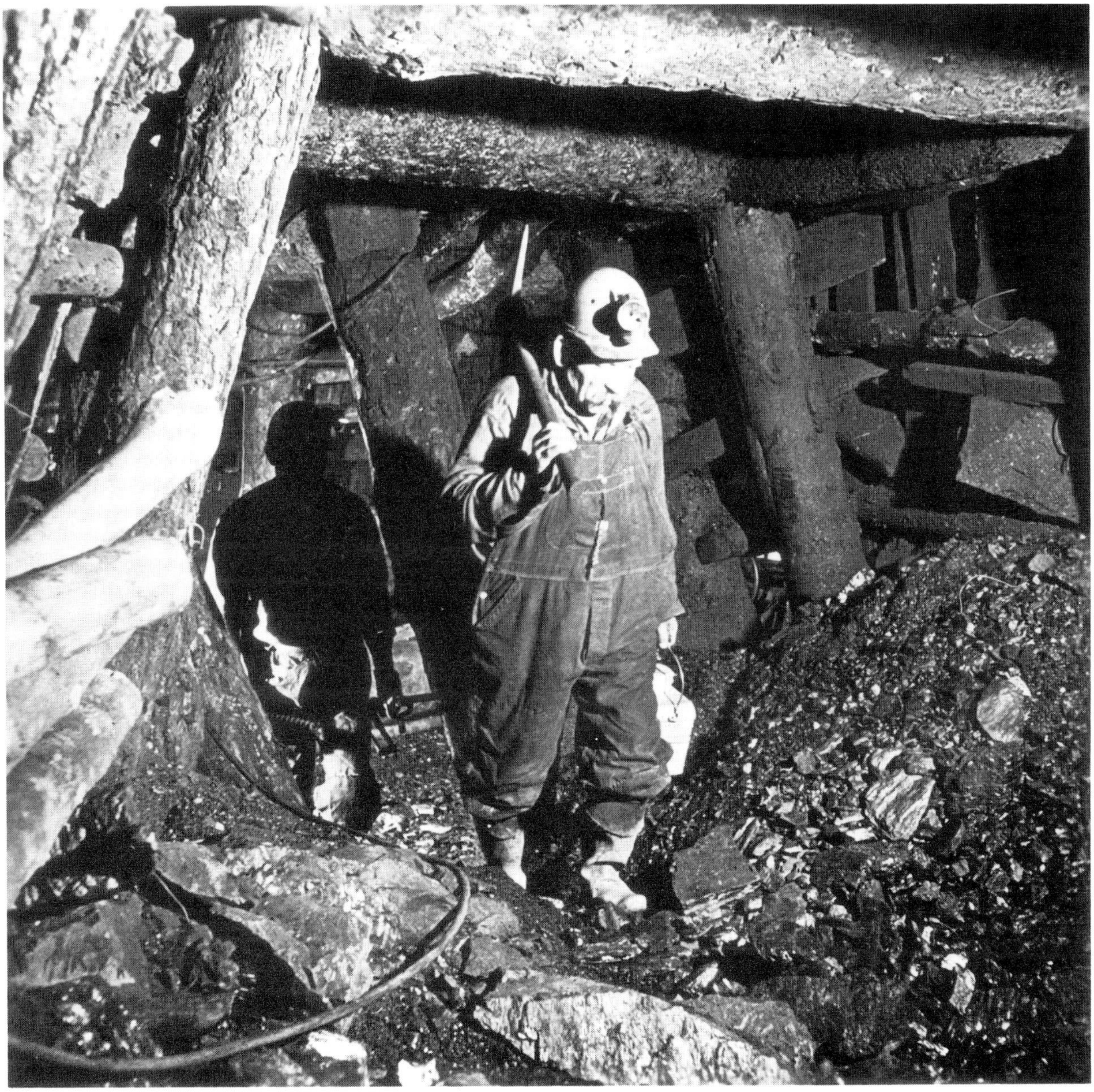

11. Andrew always precedes his helper, holds himself responsible for Mikula's safety. Here they are nearing the location for the day's work.

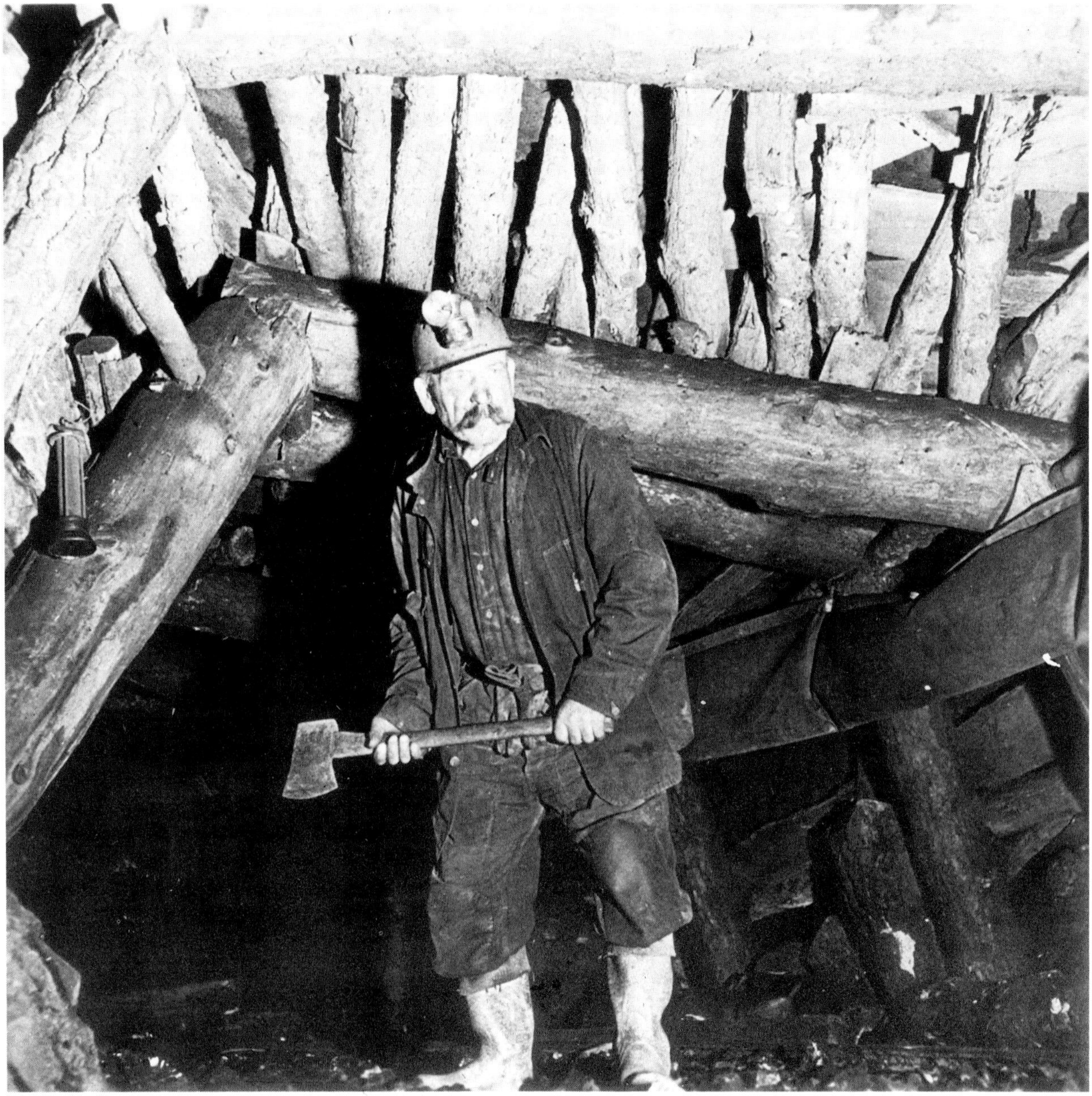

12. Andrew stops to check up on the ruggedness of timber supports. The canvas tube curving around to the right carries air.

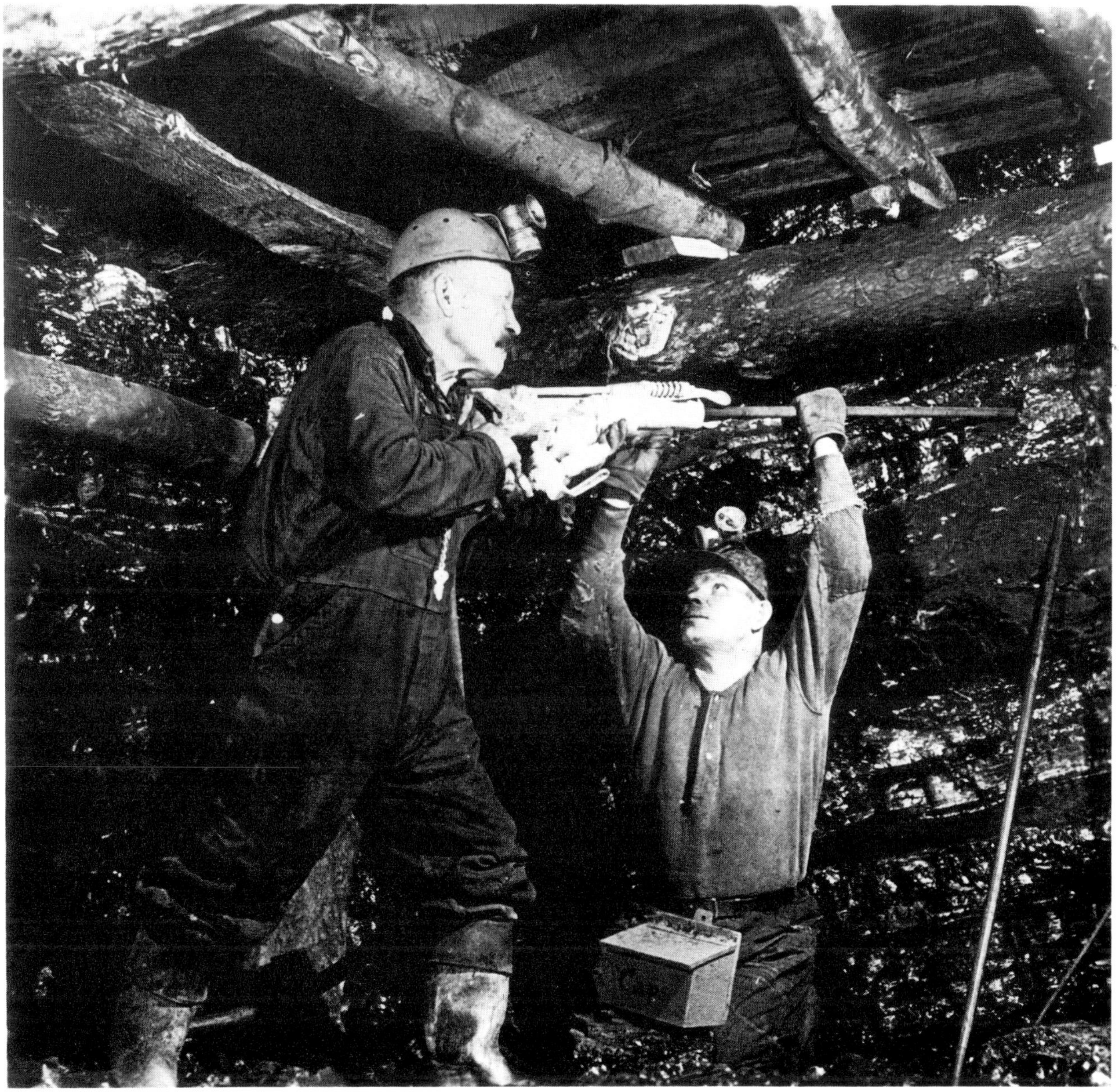

13. Andrew drills with a Jackhammer while John Mikula steadies the drill end.

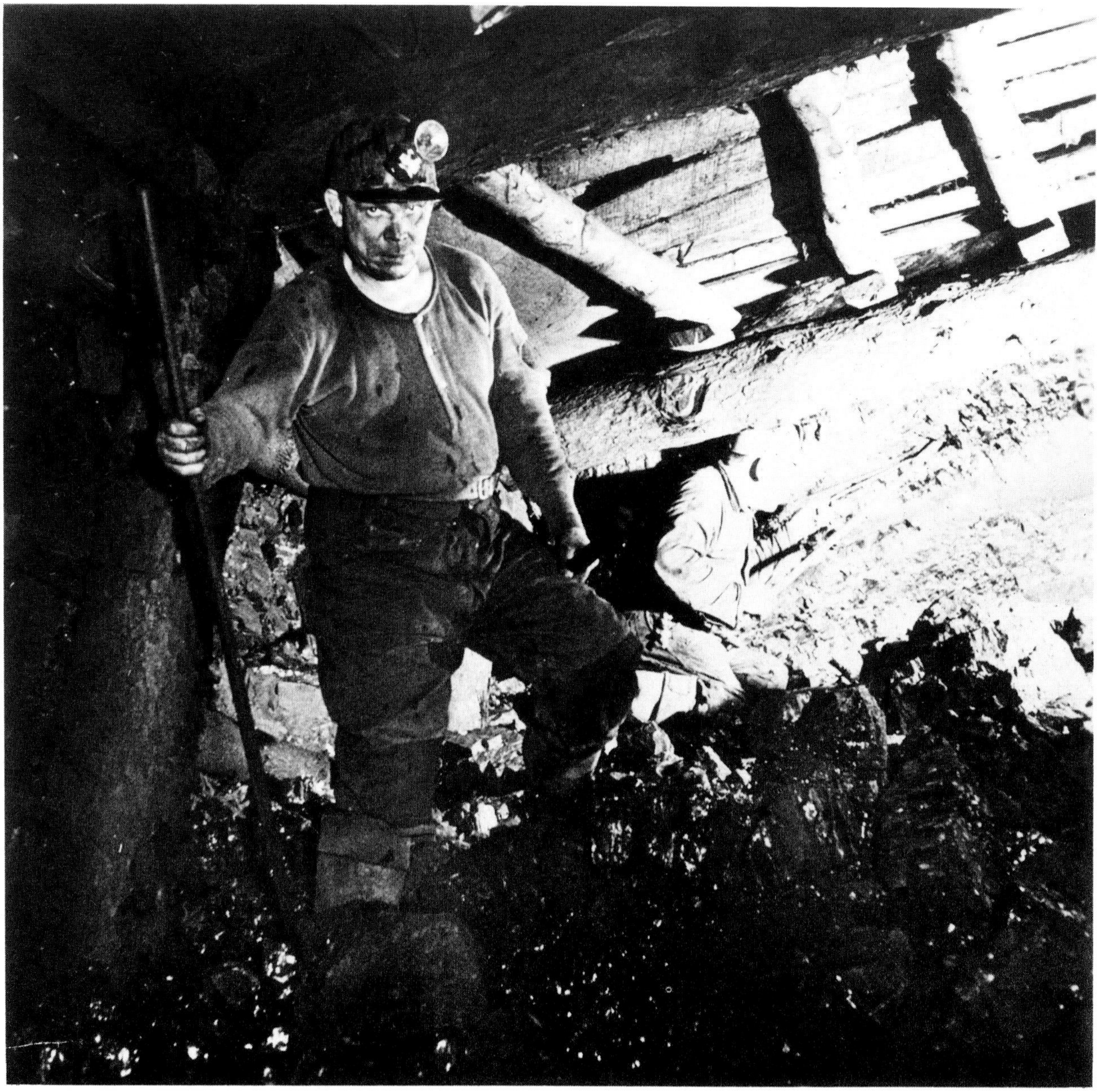

14. While Andrew places the dynamite charge and arranges the fuse he orders Mikula to stand at a safe distance.

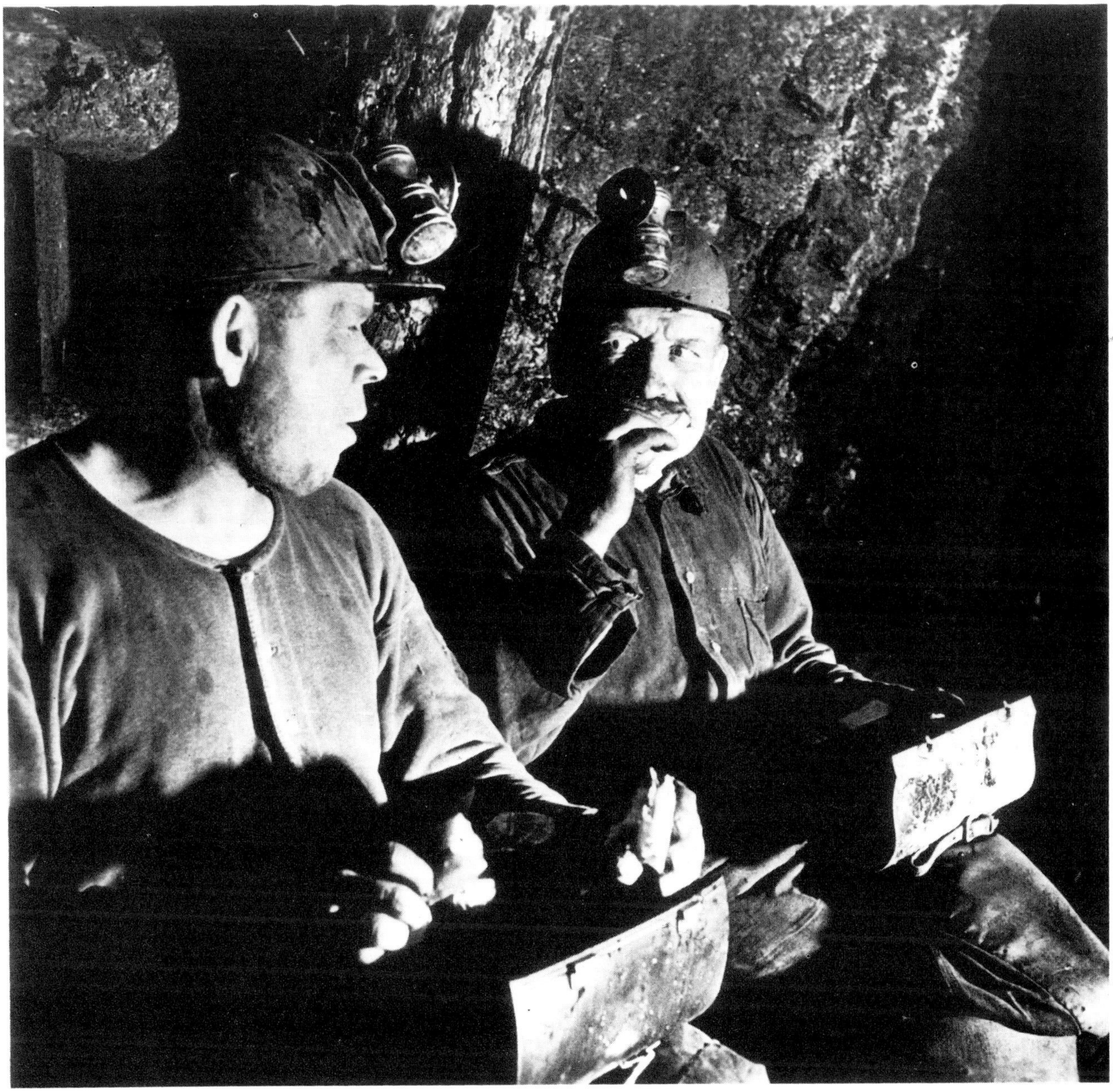

15. **After blasting loose a good pile of coal Andrew and his helper stop for lunch.**

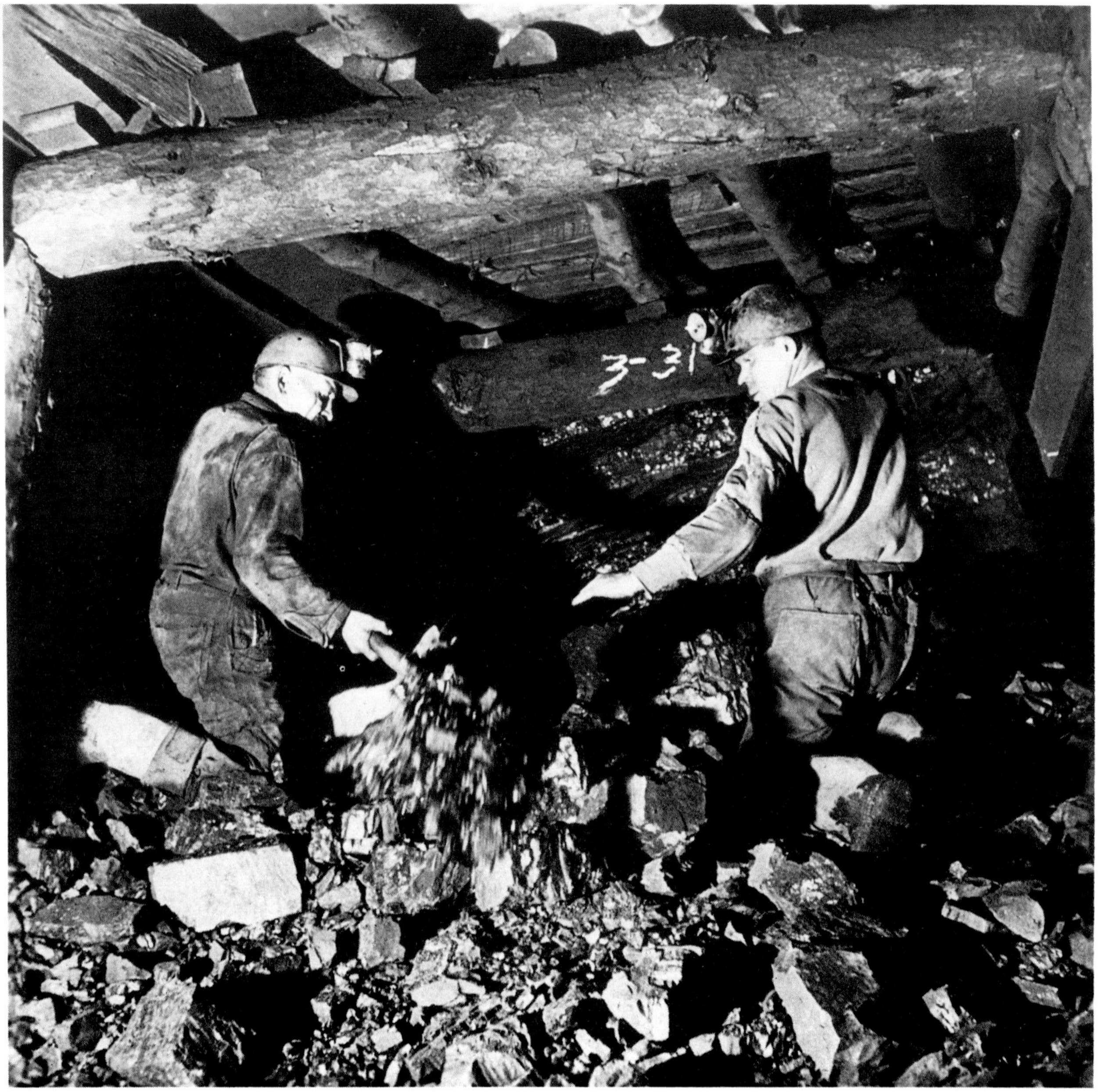

16. **Back on the job they shovel the coal out toward the buggy.**

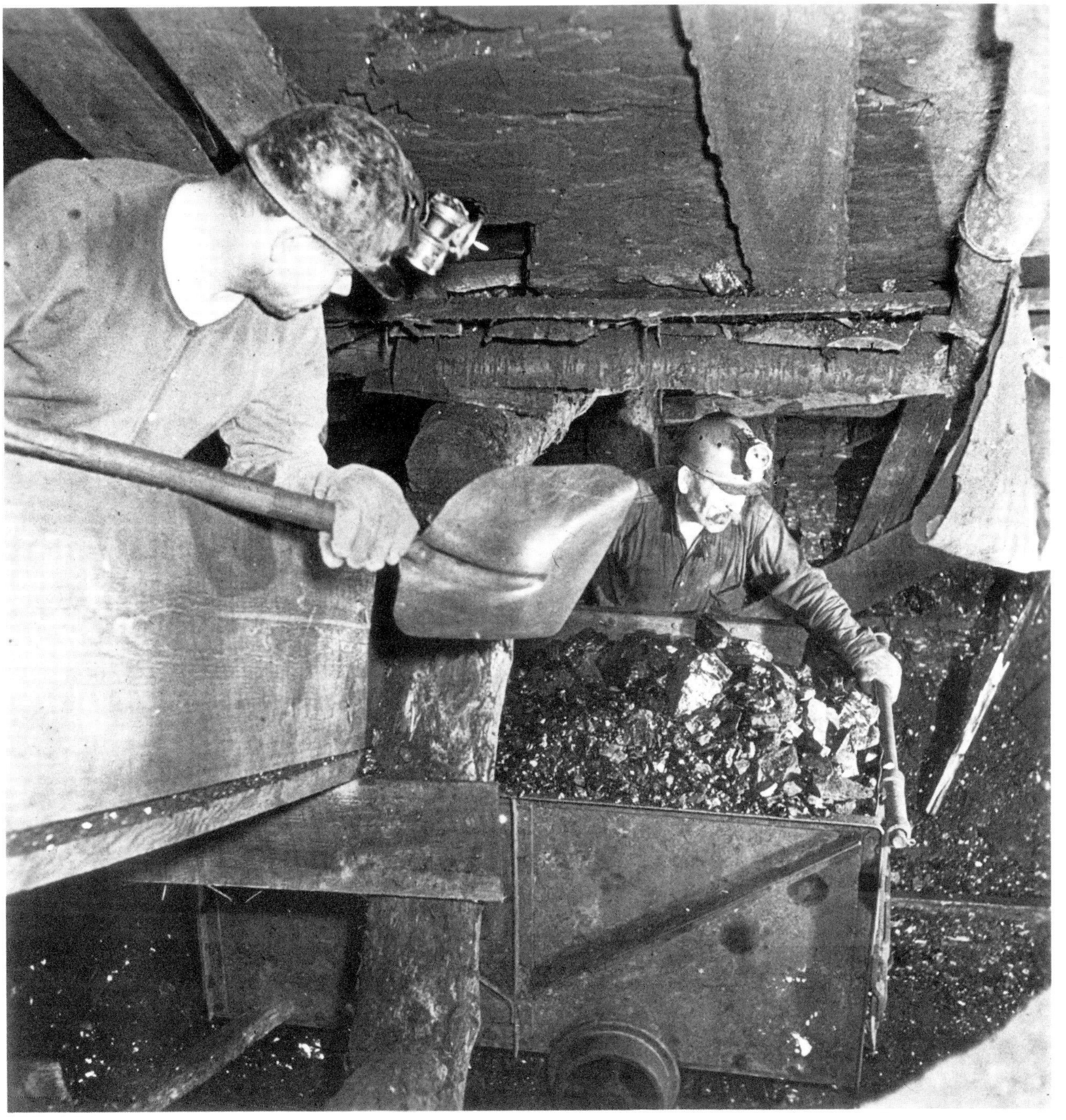

17. Andrew and Mikula have loaded the buggy with a ton and a half ...

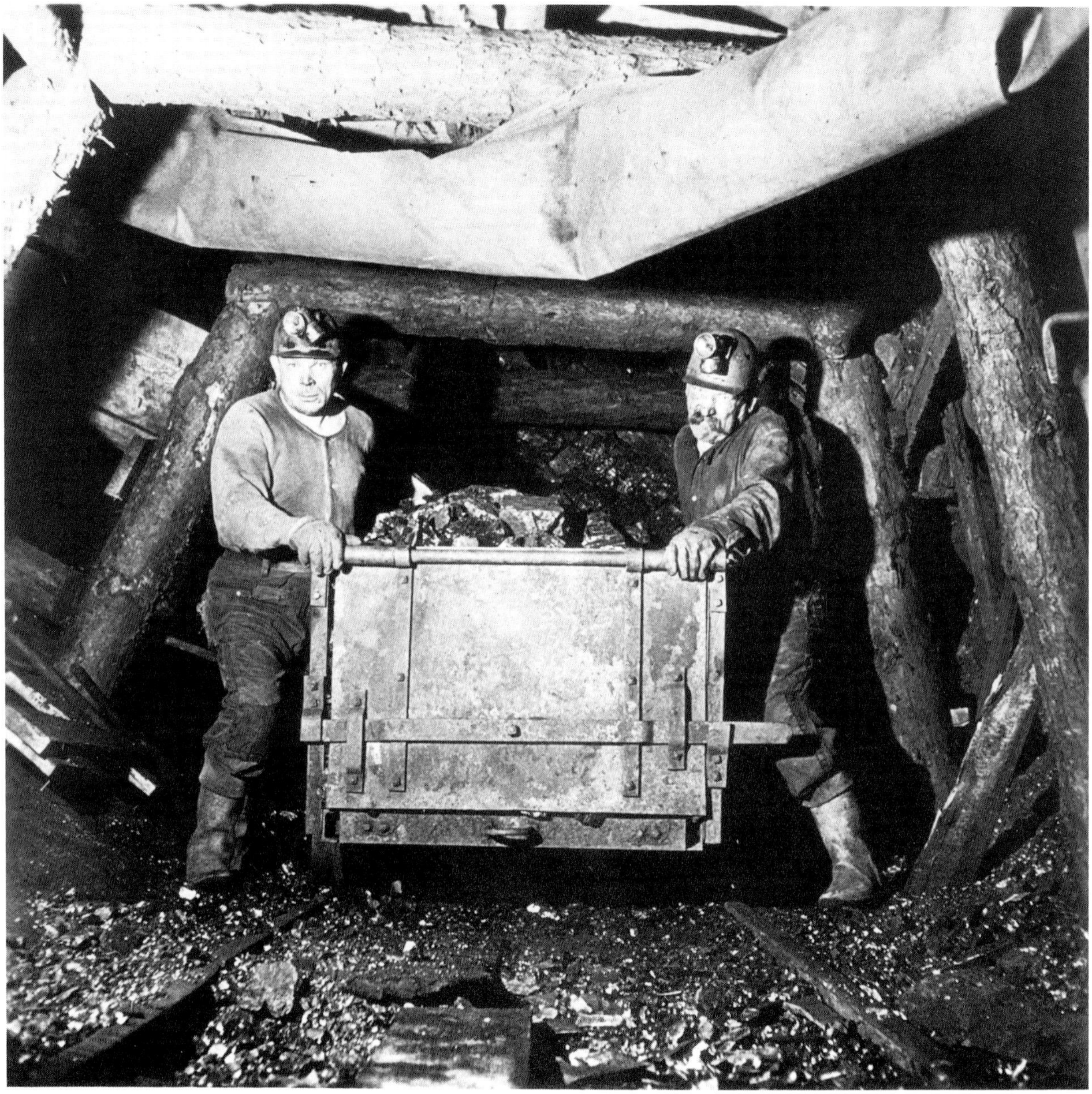

18. ... and push it out from the side vein to the main chamber.

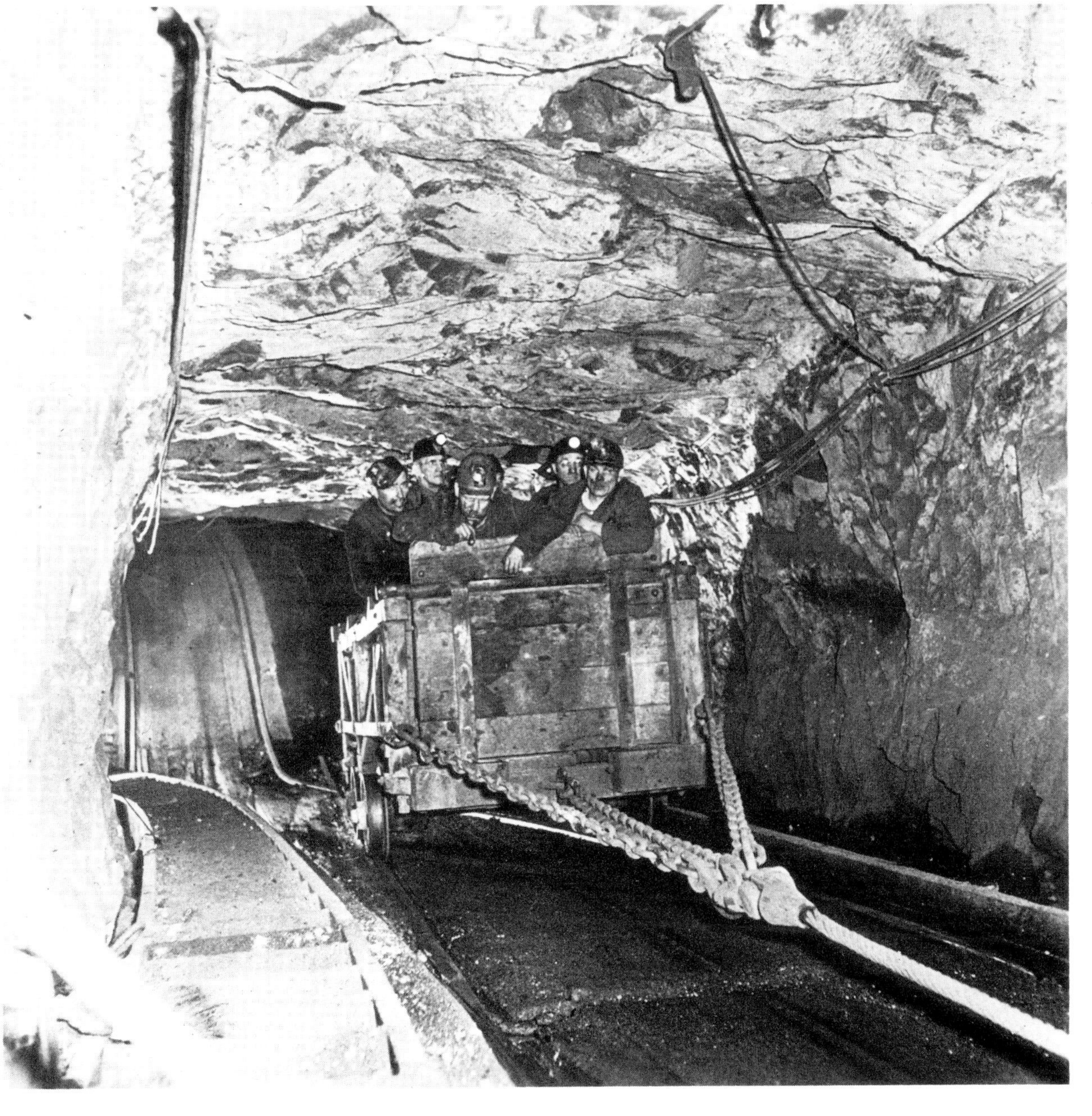

19. After the coal has been towed from the mine Andrew and Mikula join the rest of their shift-mates in the man car.

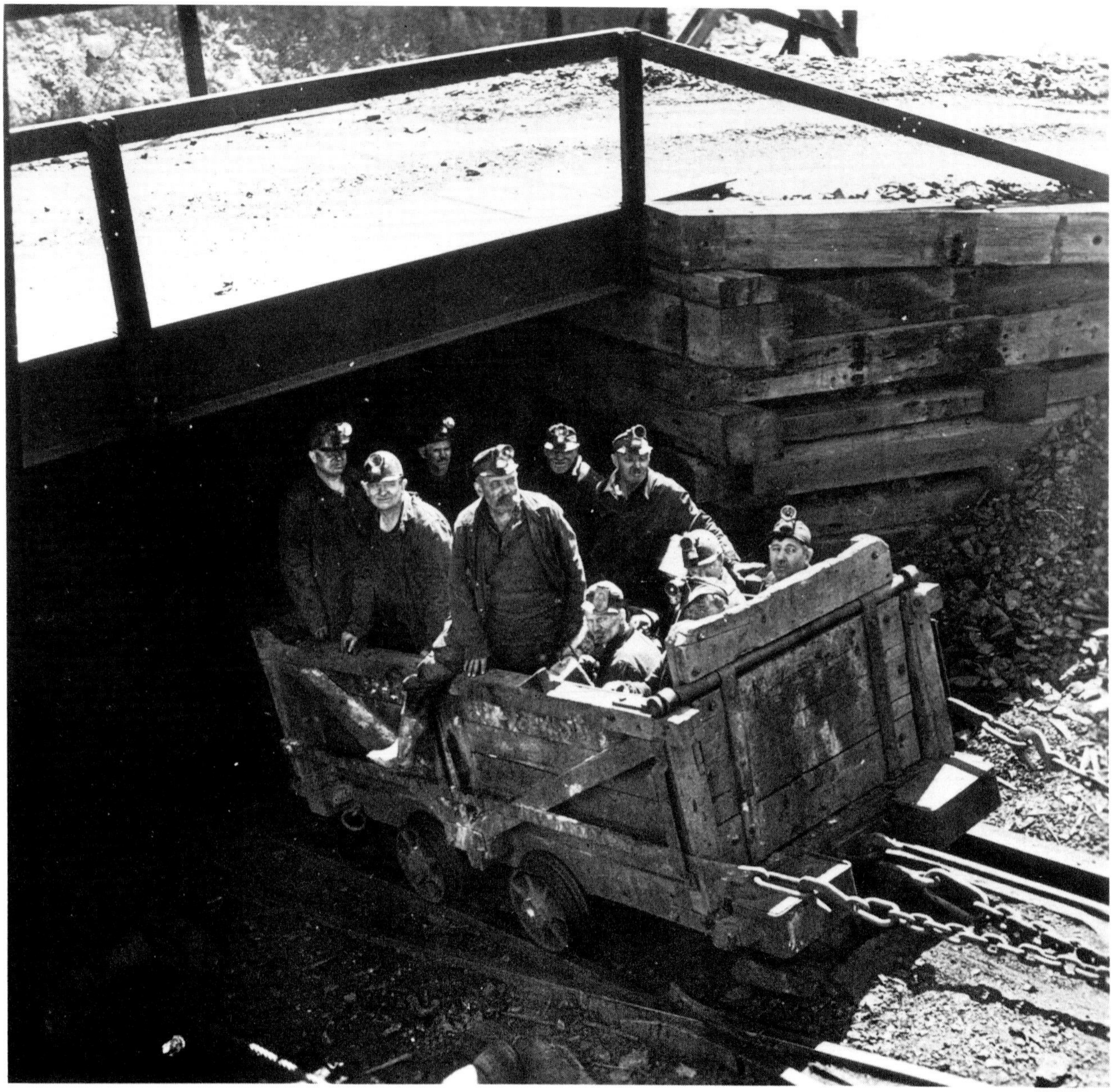

20. In the open air again, Andrew gets ready to leave the man car.

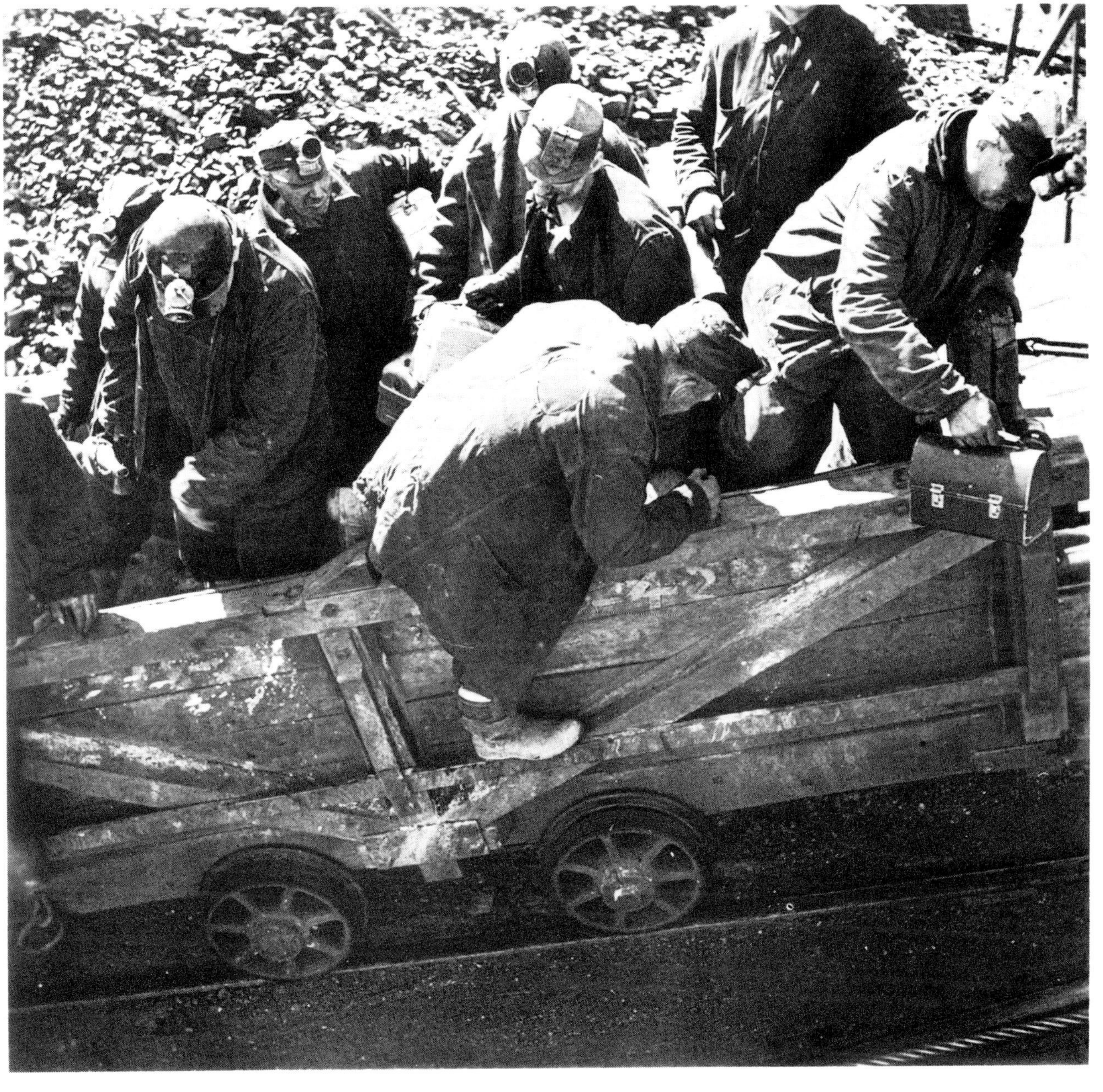

21. **They all climb out.**

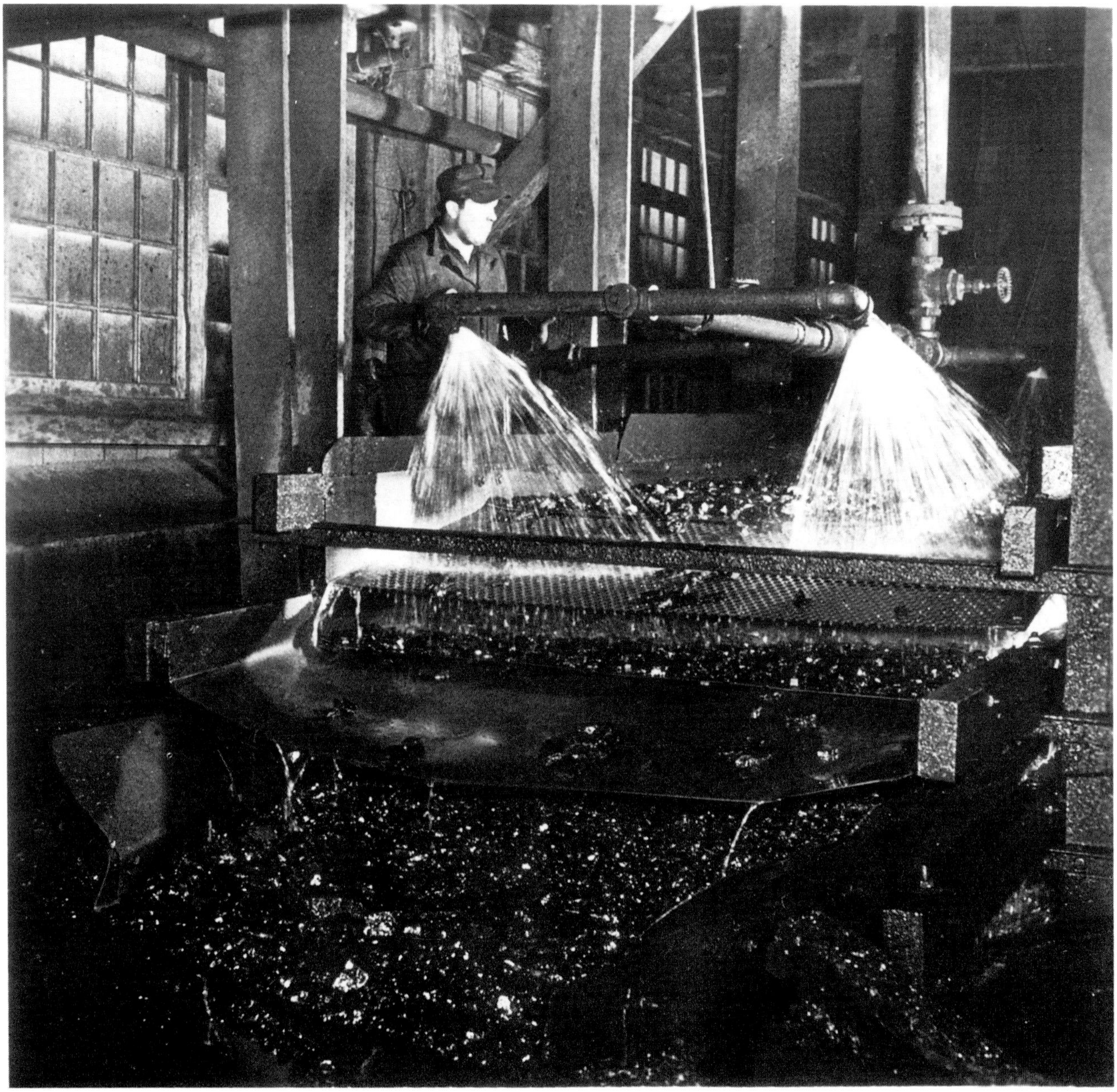

22. Meanwhile the coal is being screened and washed in the breaker. One of Andrew's sons performs this operation.

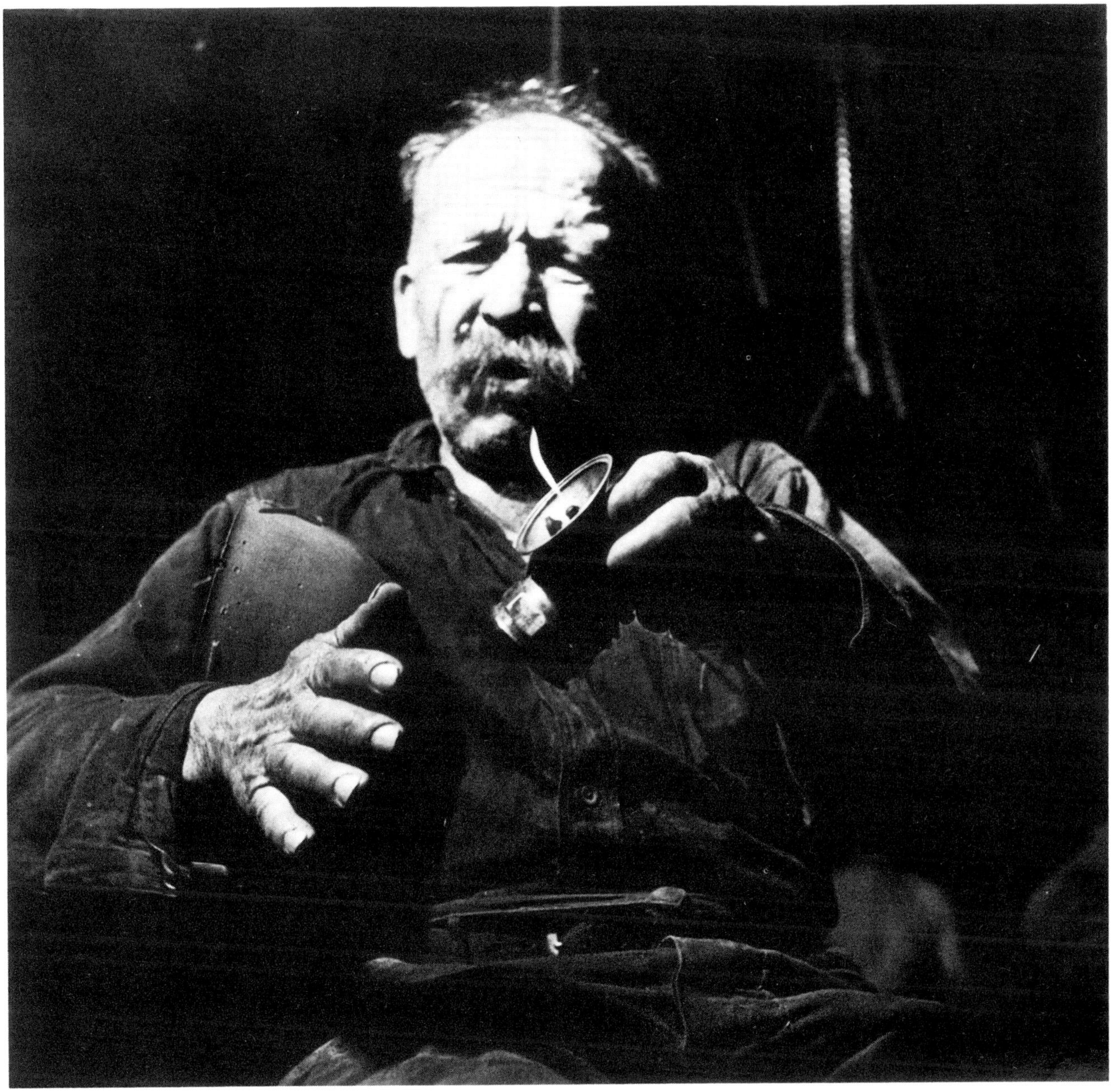

23. In the miner's dressing room after work Andrew reiterates to the photographer the importance to his safety underground of the miner's carbide lamp.

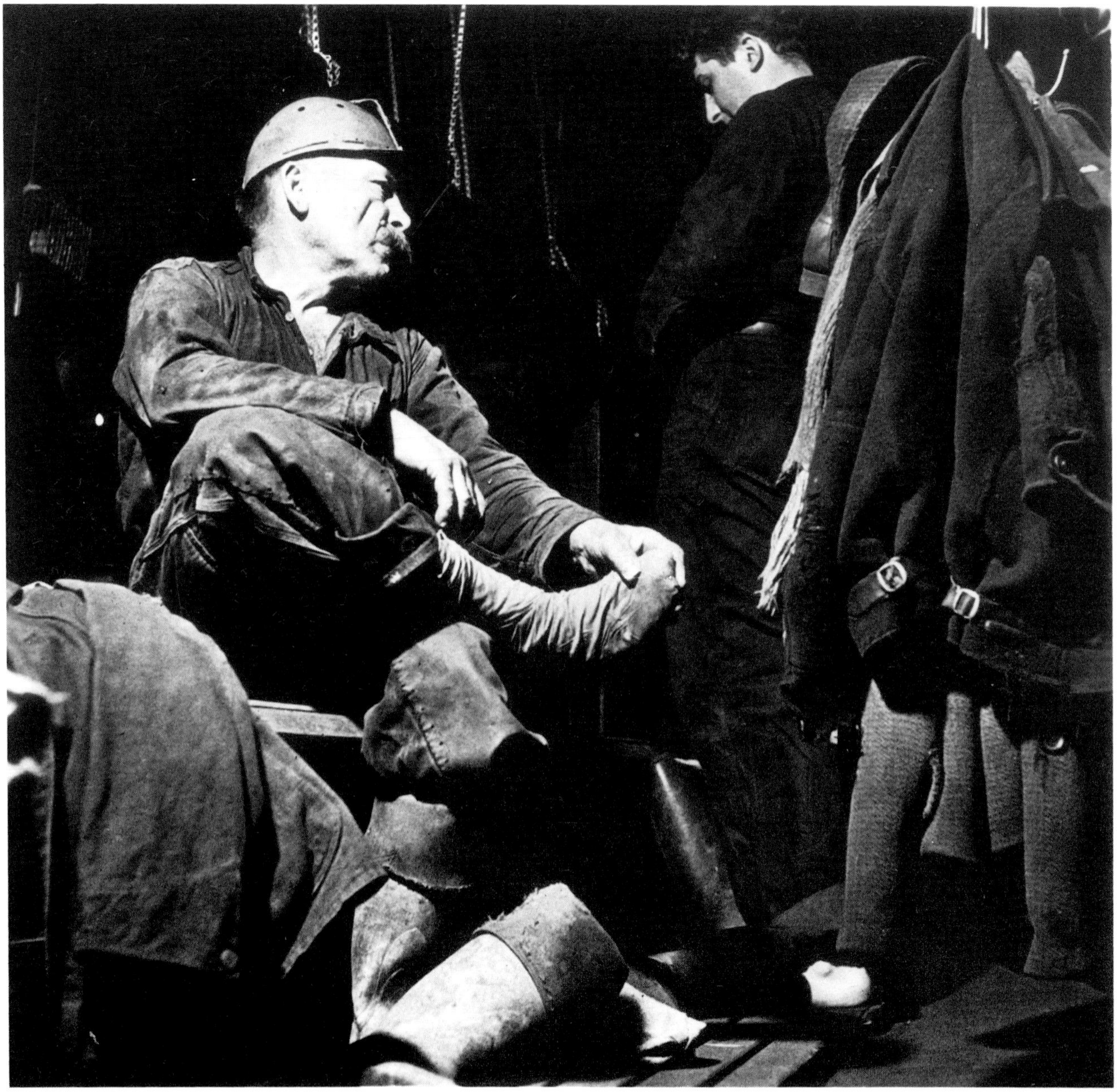

24. Having said his say Andrew starts to remove his work clothes. He will hang them on a hook which will be pulled up to dry out overnight. The dressing room is kept at a steady temperature of 85 degrees Fahrenheit.

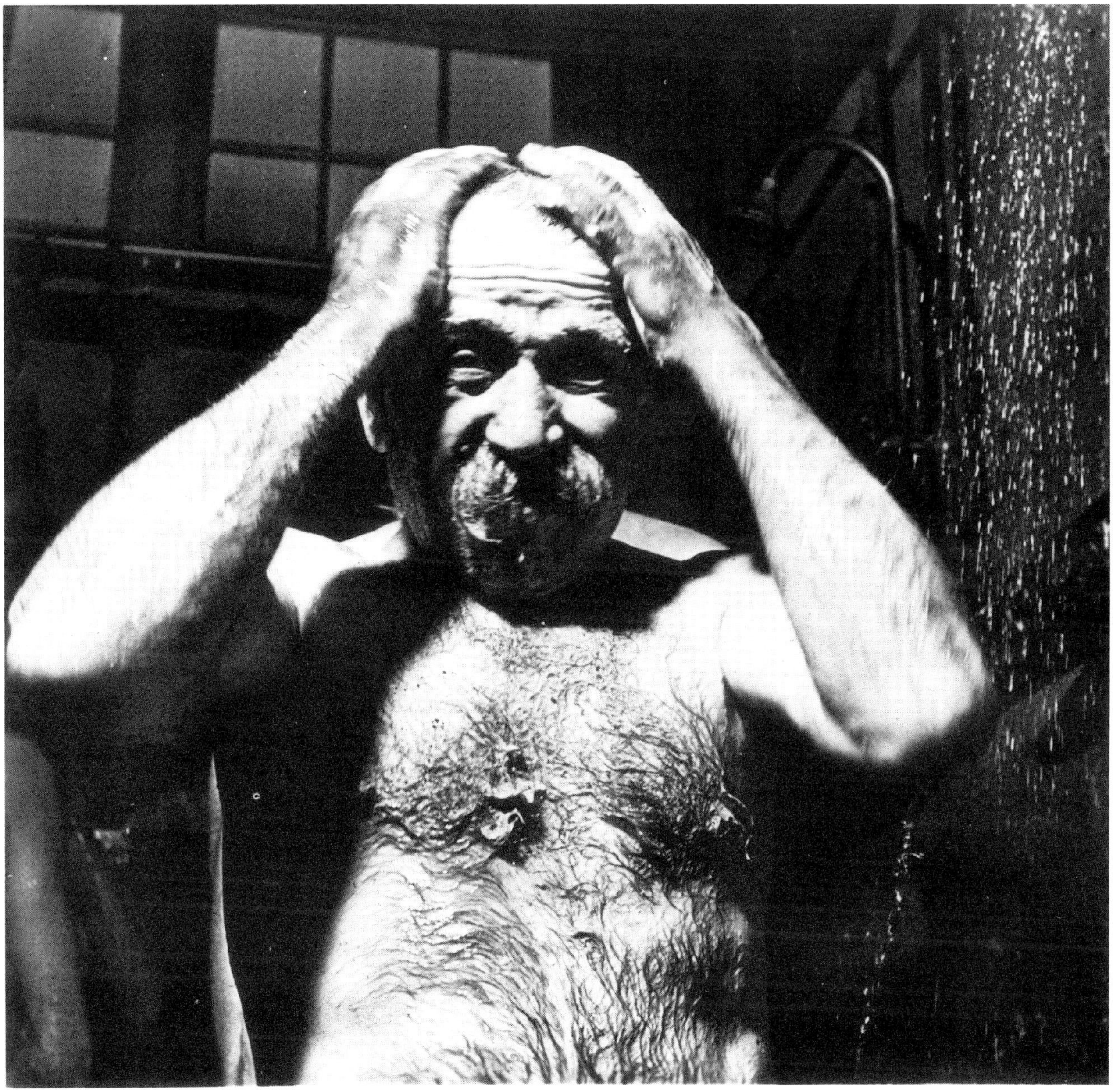

25. **Next Andrew has a good hot shower. After thirty-six years in the mine most of Andrew's hair has been rubbed off on timbers and ledges but his luxuriant moustache remains in all its vigor.**

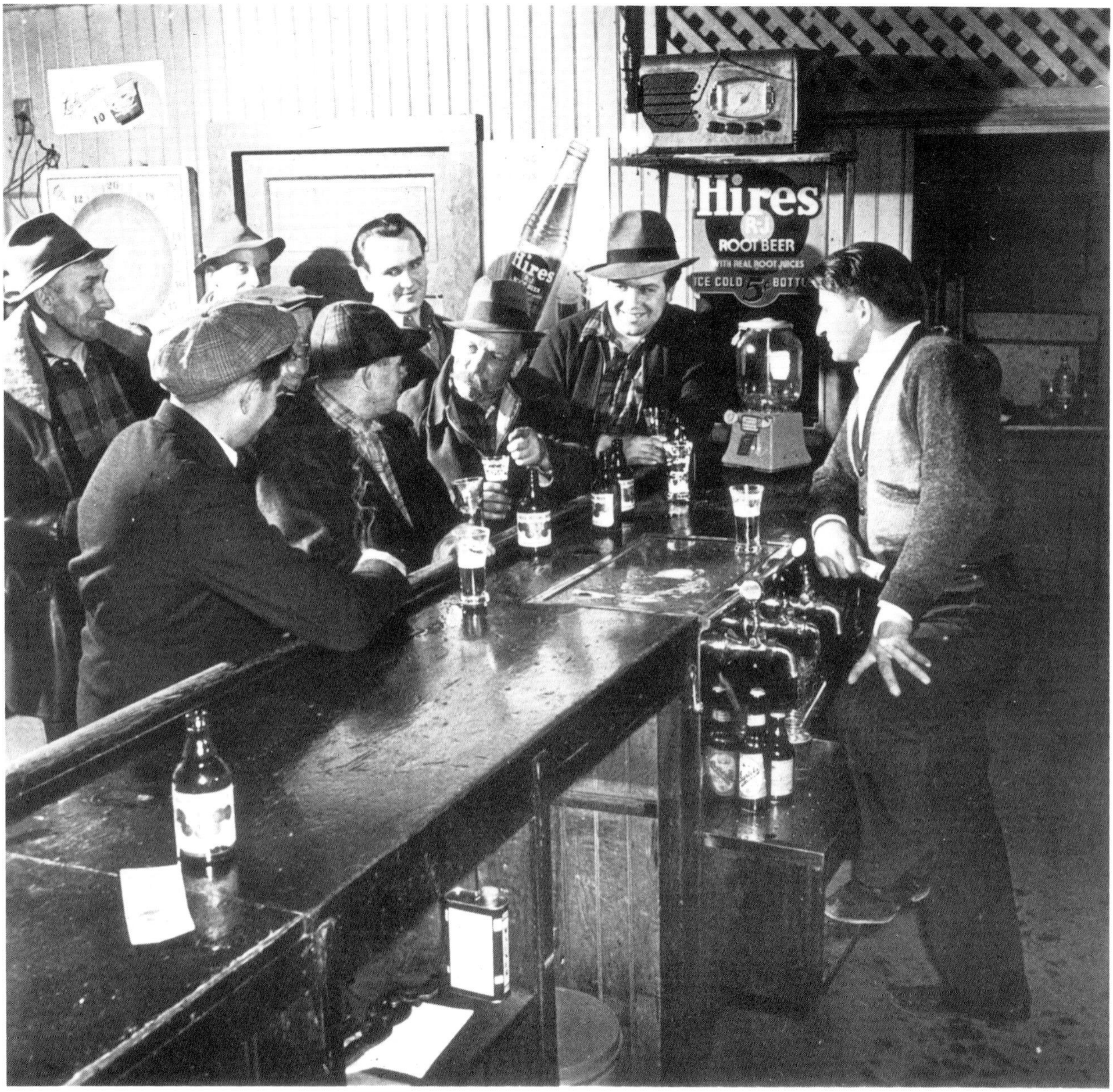

26. On the way home Andrew stops at Butch Ornifer's bar for a bottle of beer, and conversation with his neighbors. Ornifer himself is a miner who operates his bar only after working hours.

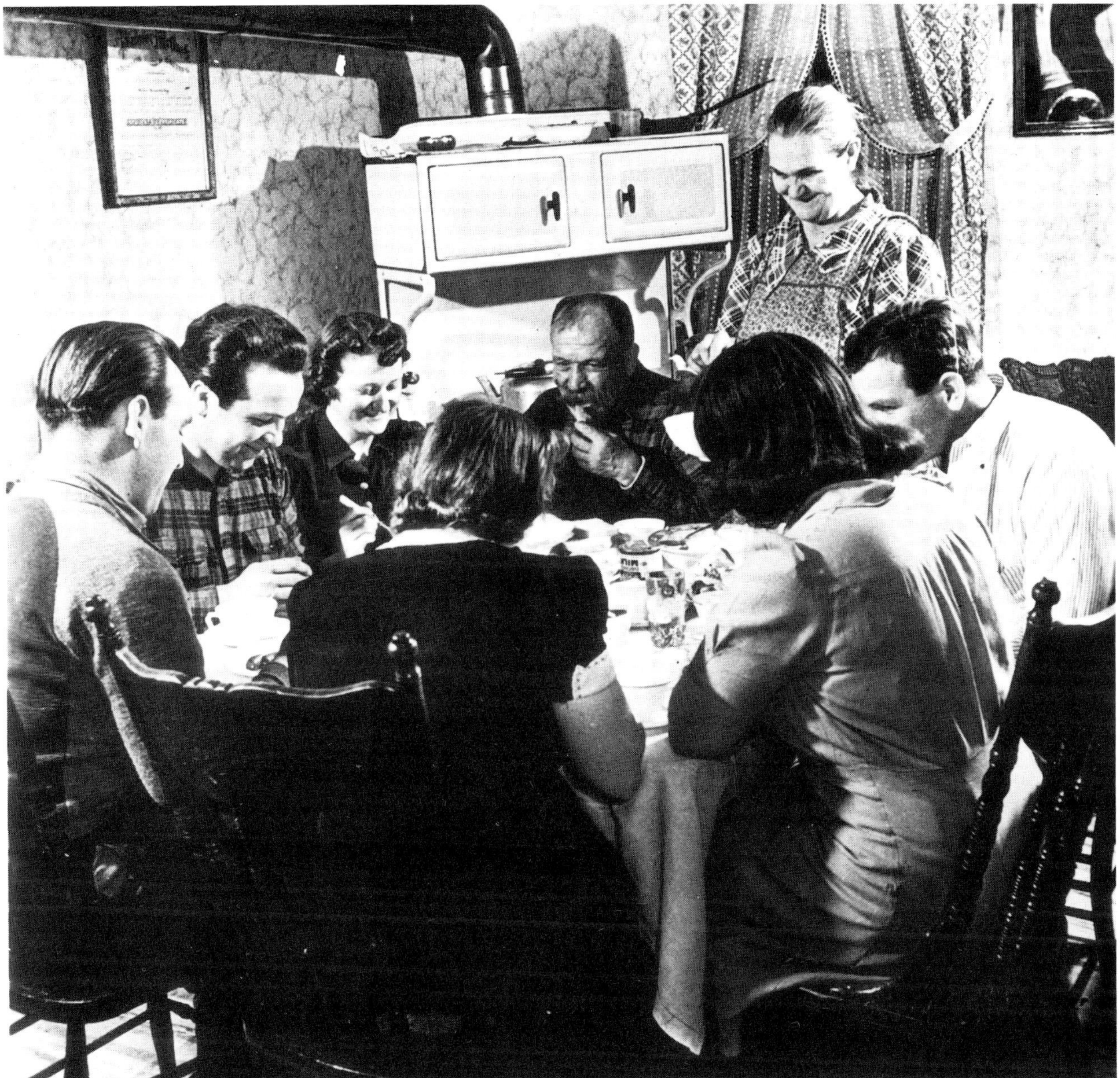

27. Family dinner at the Scavnickys revolves around the culinary expertness of Mrs. Scavnicky - Momo to her children. Andrew goes to bed soon after dinner.

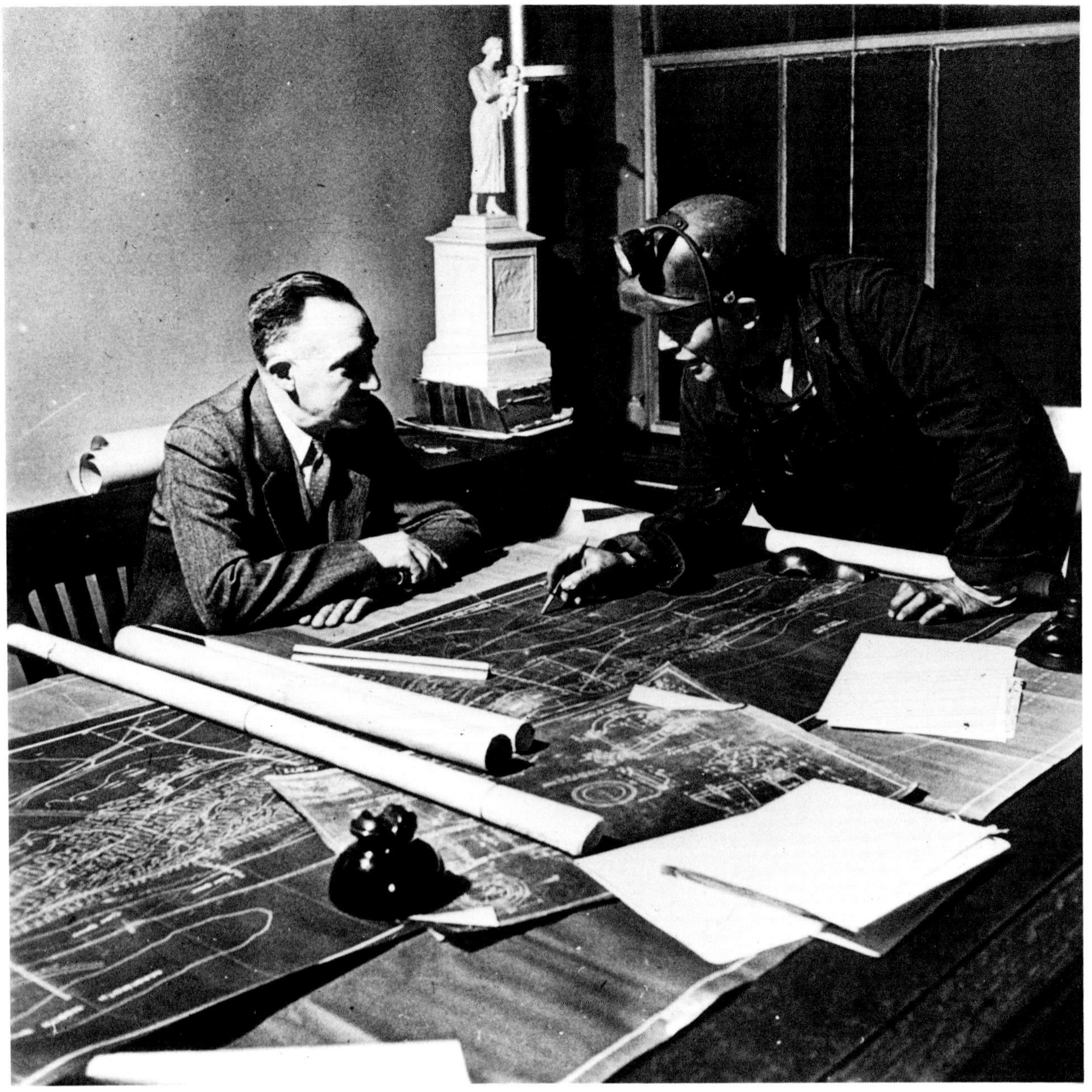

28. Andrew and other miners at Jeddo-Highland Mine No. 7 benefit from the excellent safety record established there. Here Andrew's immediate superior, a mine foreman, reports on conditions underground to an officer of the company. The figure in the corner by the window is a safety trophy won by the company.

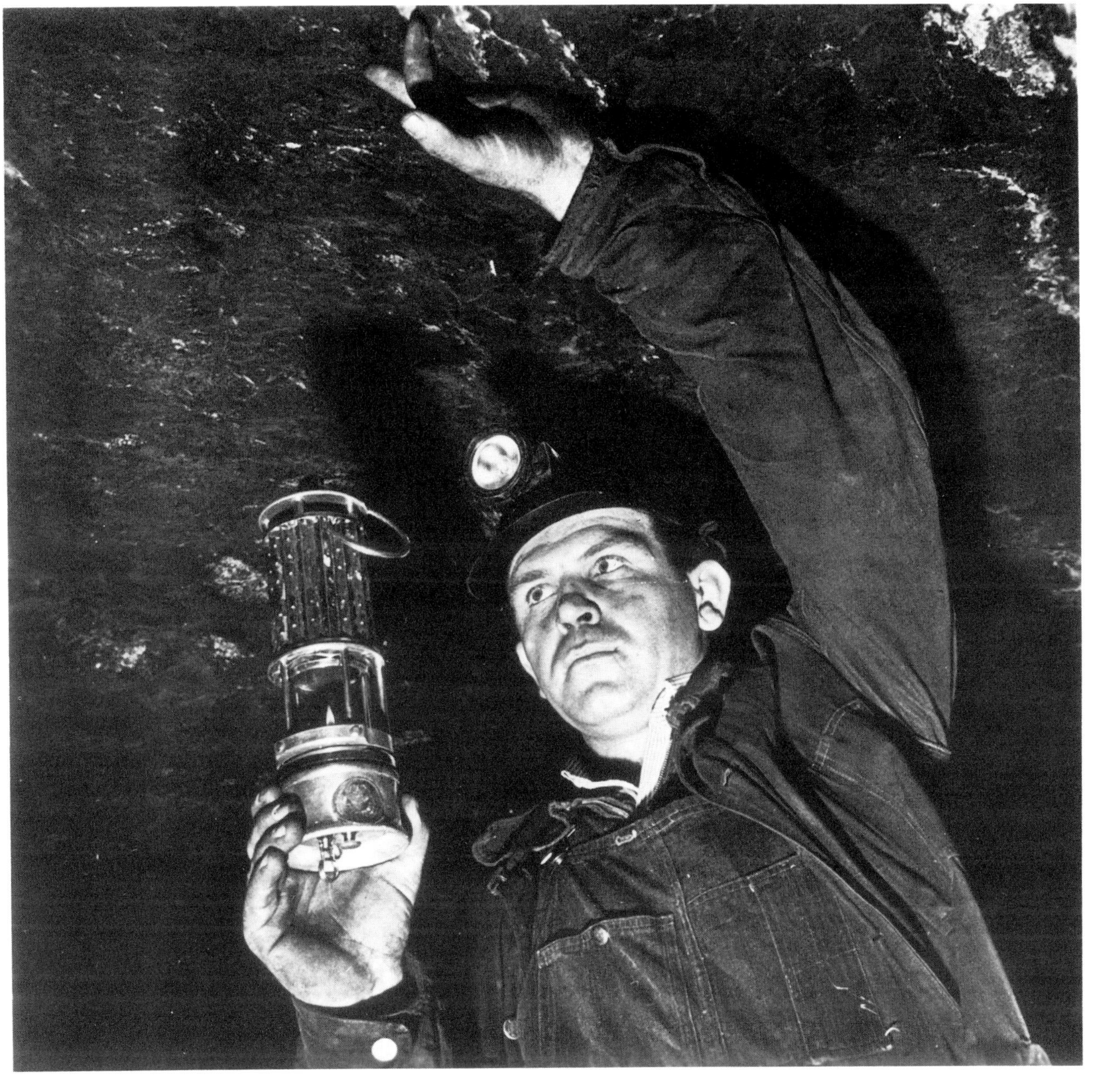

29. Another one of Andrew's sons works in the mines as an engineer, not a miner. It is his highly technical and important job to check each vein for gases and fissures in the strata. His work makes for safety and efficiency.

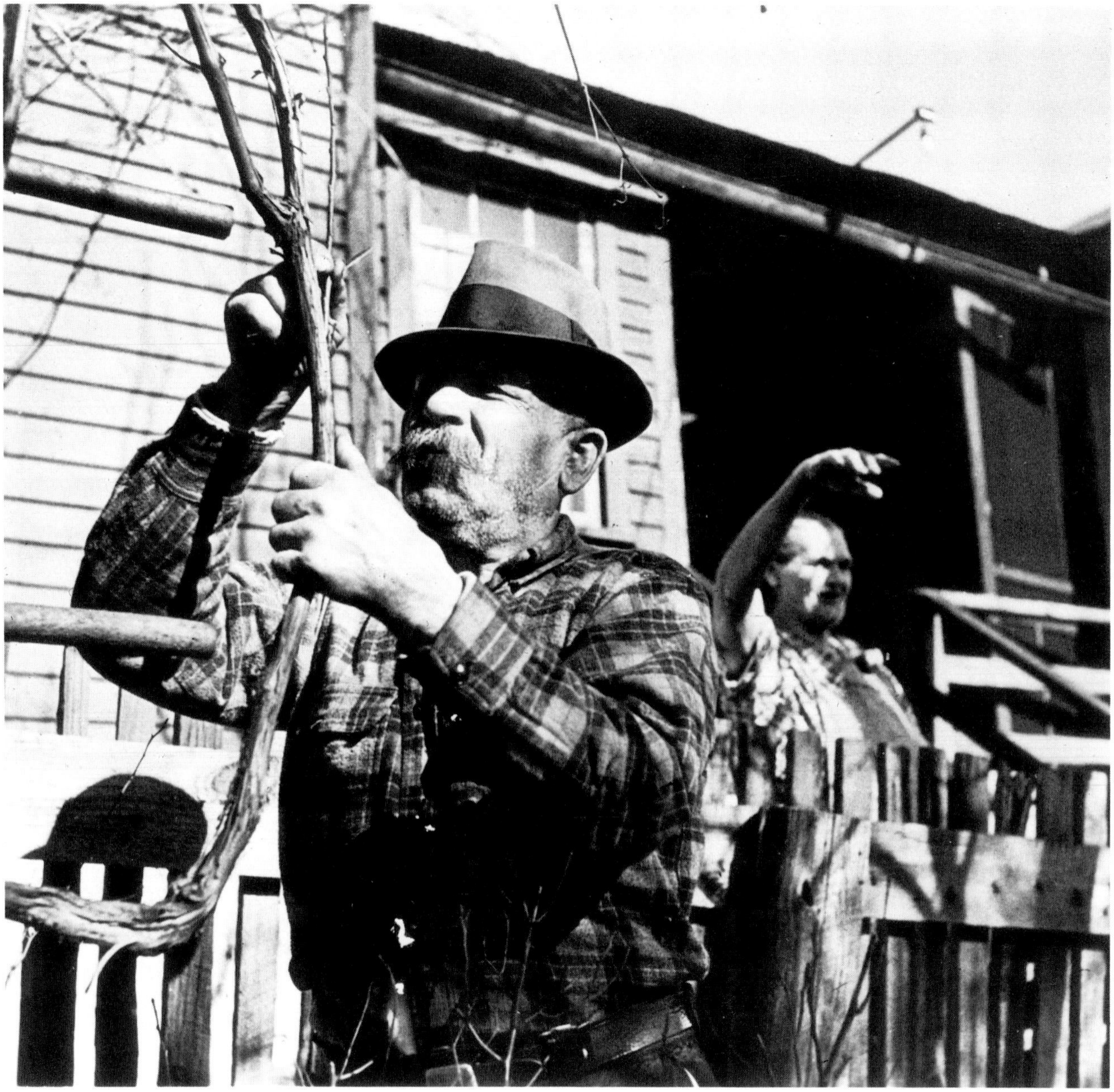

30. On Sundays, Andrew has time to follow his hobbies. Here he is shown working on a grape vine. He has considerable skill in grafting, despite lack of training. He has developed one tree which bears apples, cherries, pears and plums; another which bears twelve different kinds of pears.

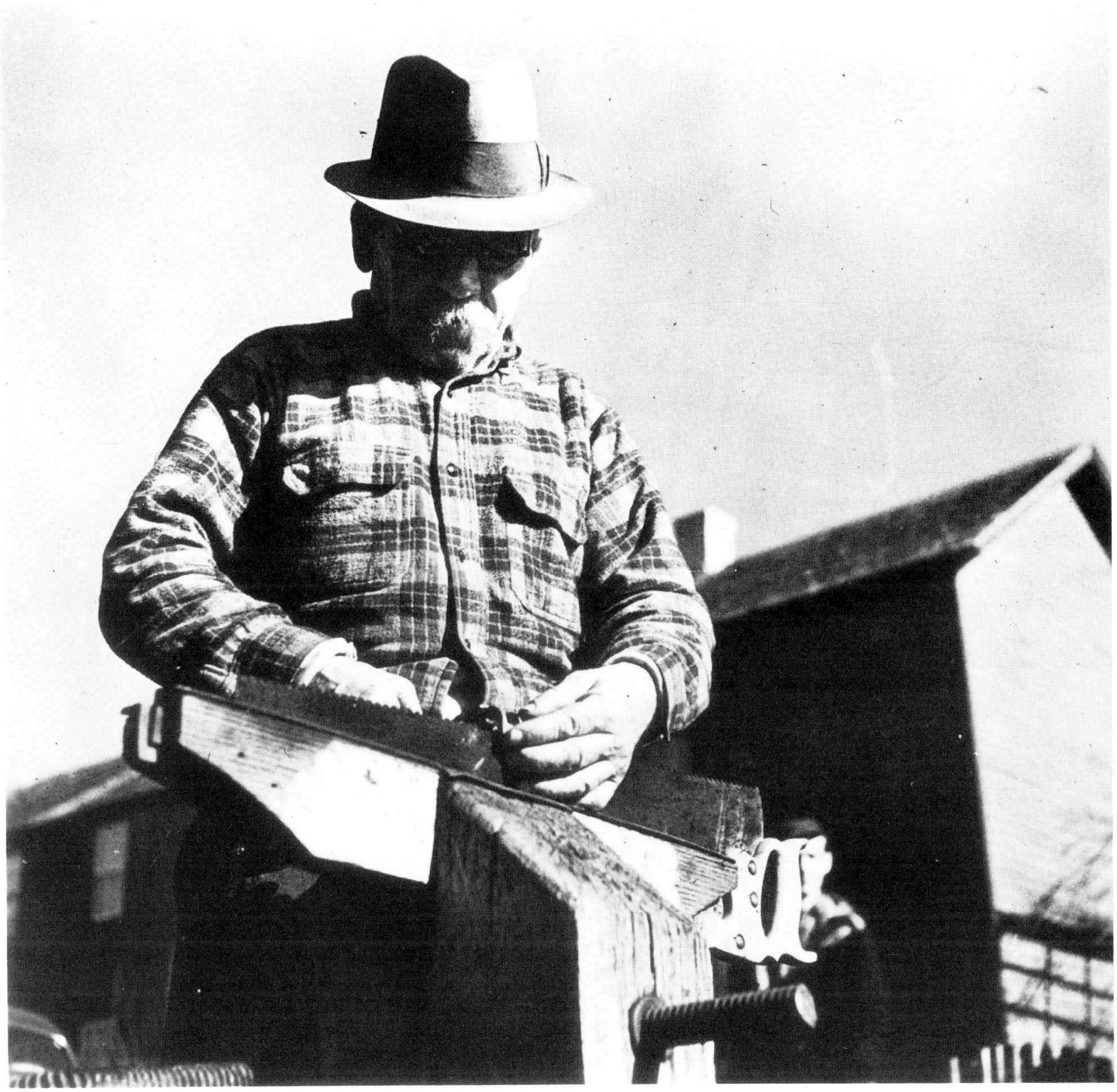

31. A miner gets to be handy with all kinds of tools. Here on a bright Sunday morning Andrew is sharpening a saw.

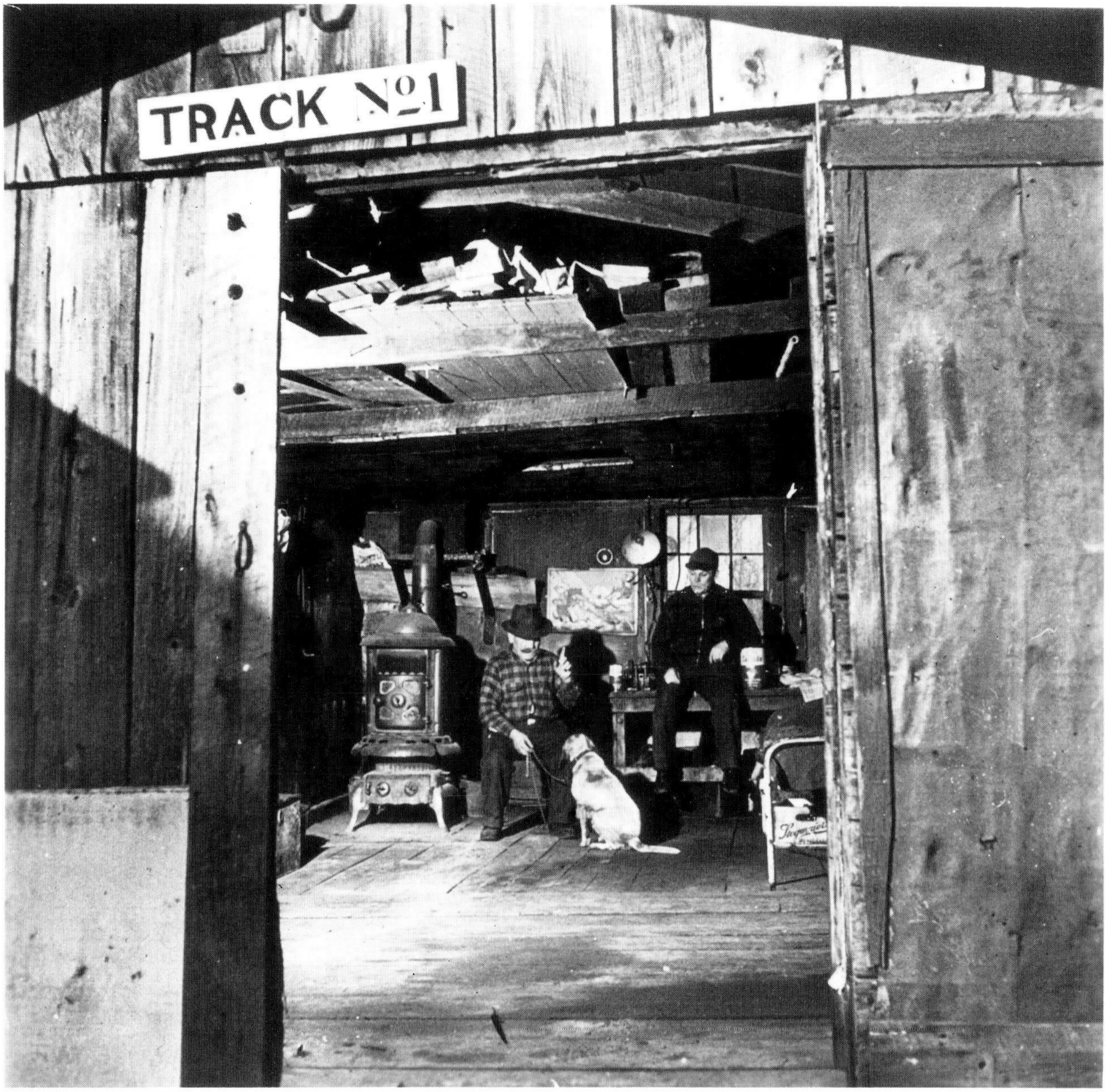

32. During off hours Andrew likes to sit in his workshop, pottering with tools, teaching his dog tricks, and talking to his buddy John Mikula. Andrew has swiped a sign from the mine and tacked it over the shed door.

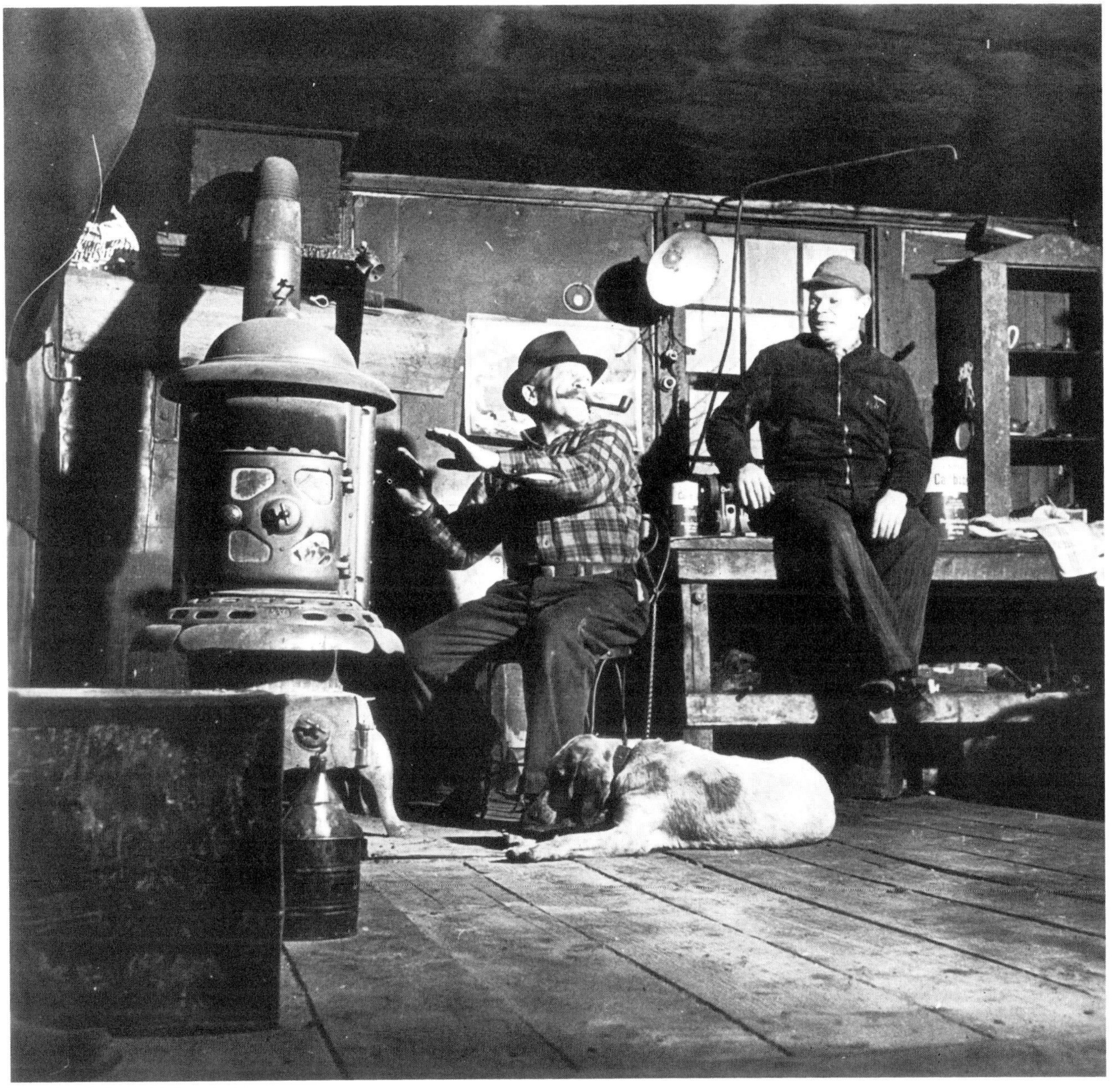

33. A workshop is also a good place to loaf. A coal stove keeps it warm even on cold days (interior of picture #32).

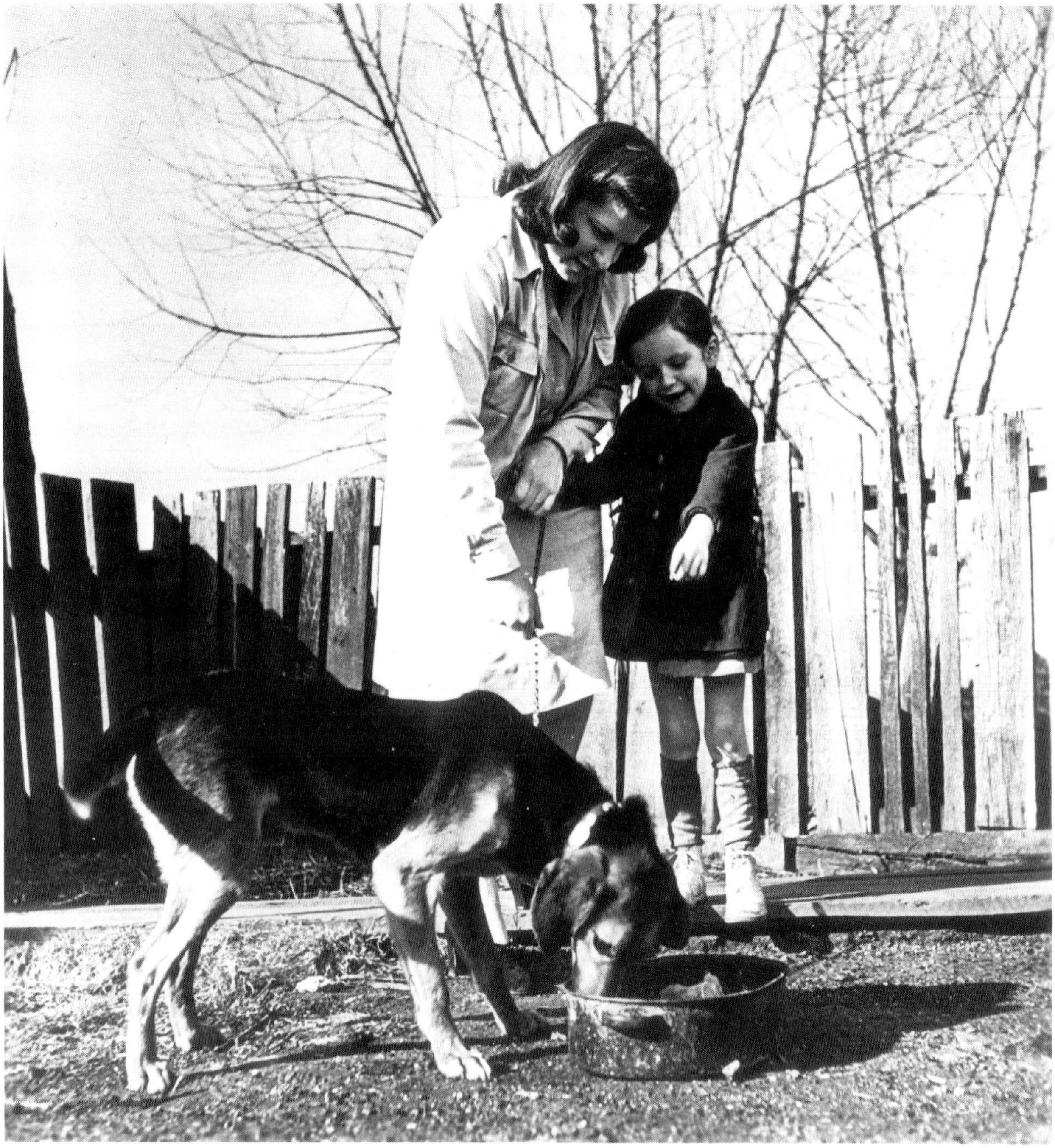

34. The Scavnickys are a companionable family. One of Andrew's daughters and granddaughters pay a Sunday call with their hound. All are welcome.

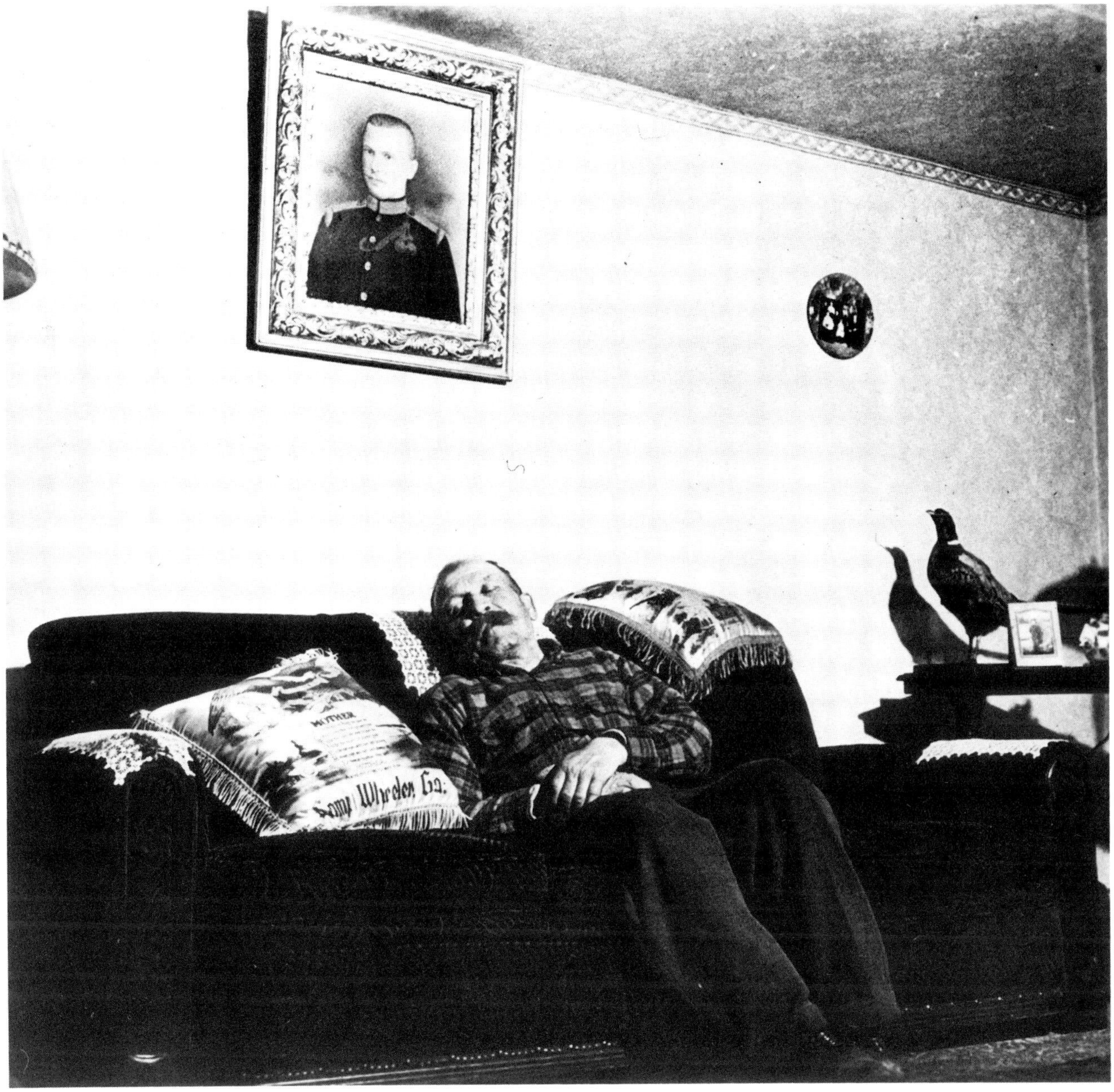

35. After a filling Sunday dinner at mid-day, Andrew likes to take a nap. After all, Sunday is the day of rest. Here he slumbers on the livingroom sofa. Above him hangs a picture of the young Andrew taken 46 years ago when he was a private in the army of Emperor Franz Josef who then ruled over Czechoslovakia. Pillow was sent home by son Michael, who is a private in the U.S. Army.

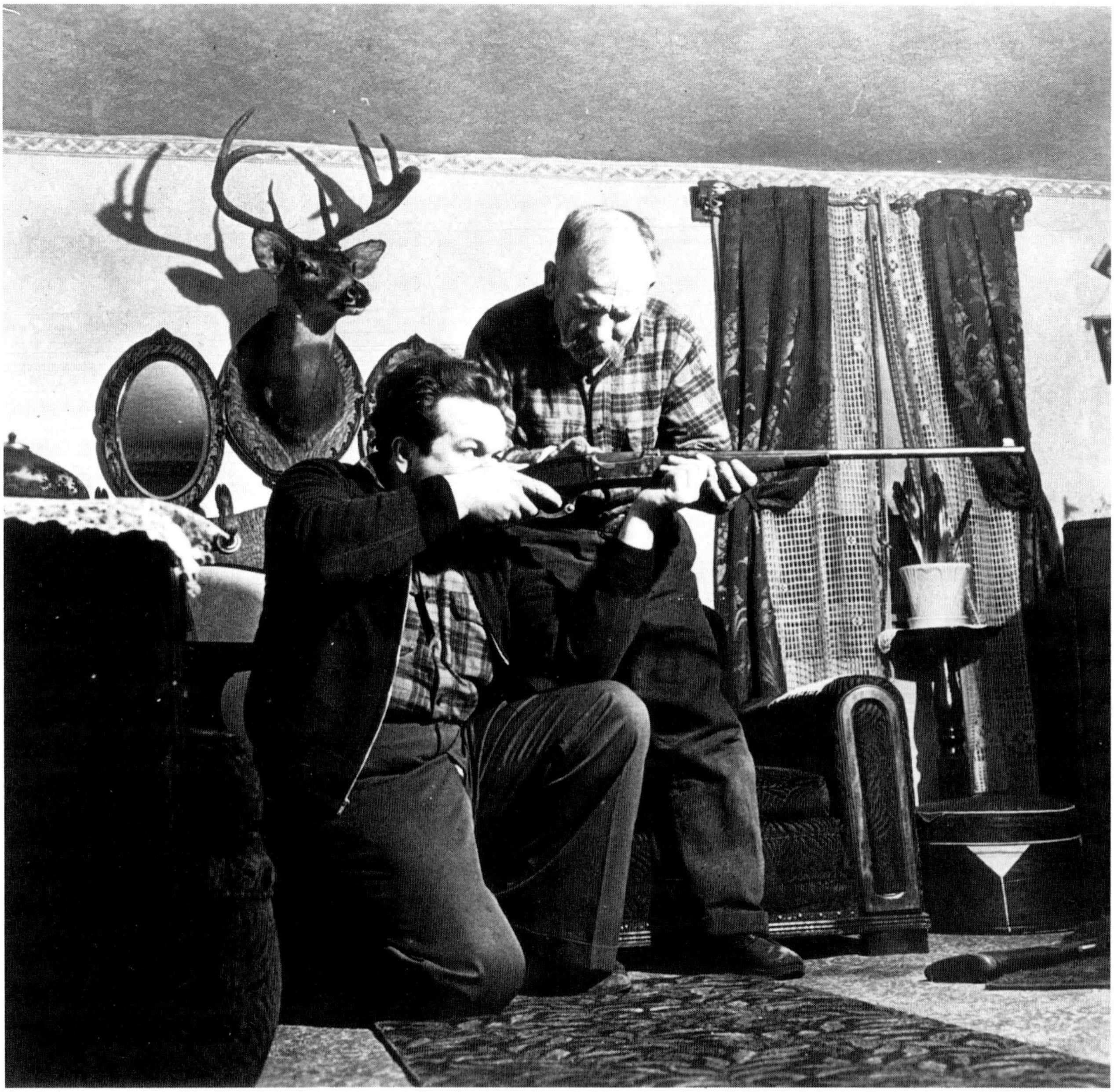

36. During his service in the Austro-Hungarian army Andrew Scavnicky won several sharpshooting medals. He is still an expert shot. And he has taught his sons respect for good rifles and skill in their use. His son Paul, here getting a little post-graduate instruction from his father, shot the buck whose head is shown hanging on the wall. Andrew is a member of the Hazleton Rifle and Revolver Club, has won medals there, too, for his prowess.

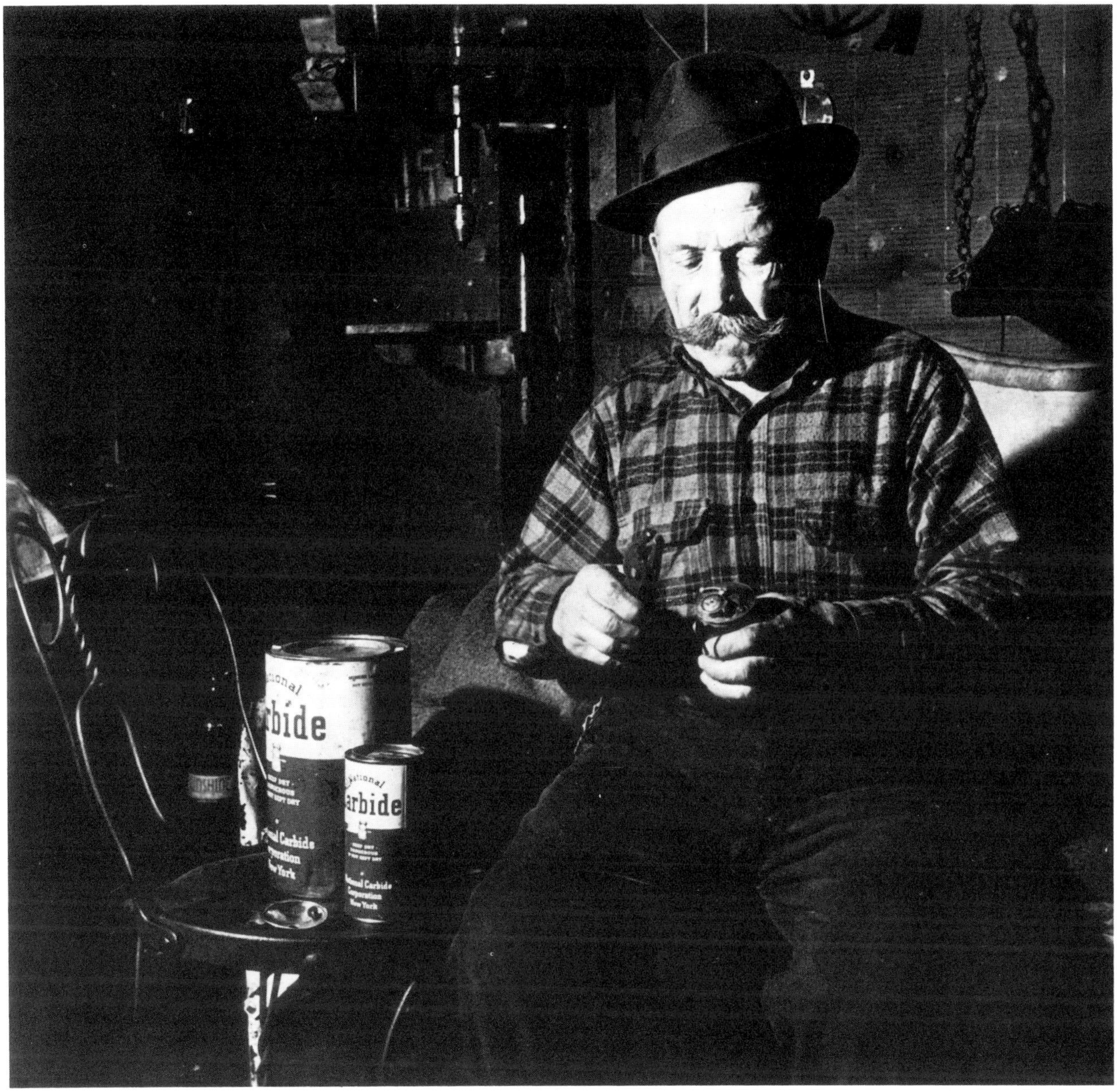

37. Wearing his old clothes but one of his two good hats, Andrew sits in his workshop Sunday evening filling and adjusting the carbide lamp which will light his labors on the morrow. Not even on off time does a miner forget the importance of his lamp.

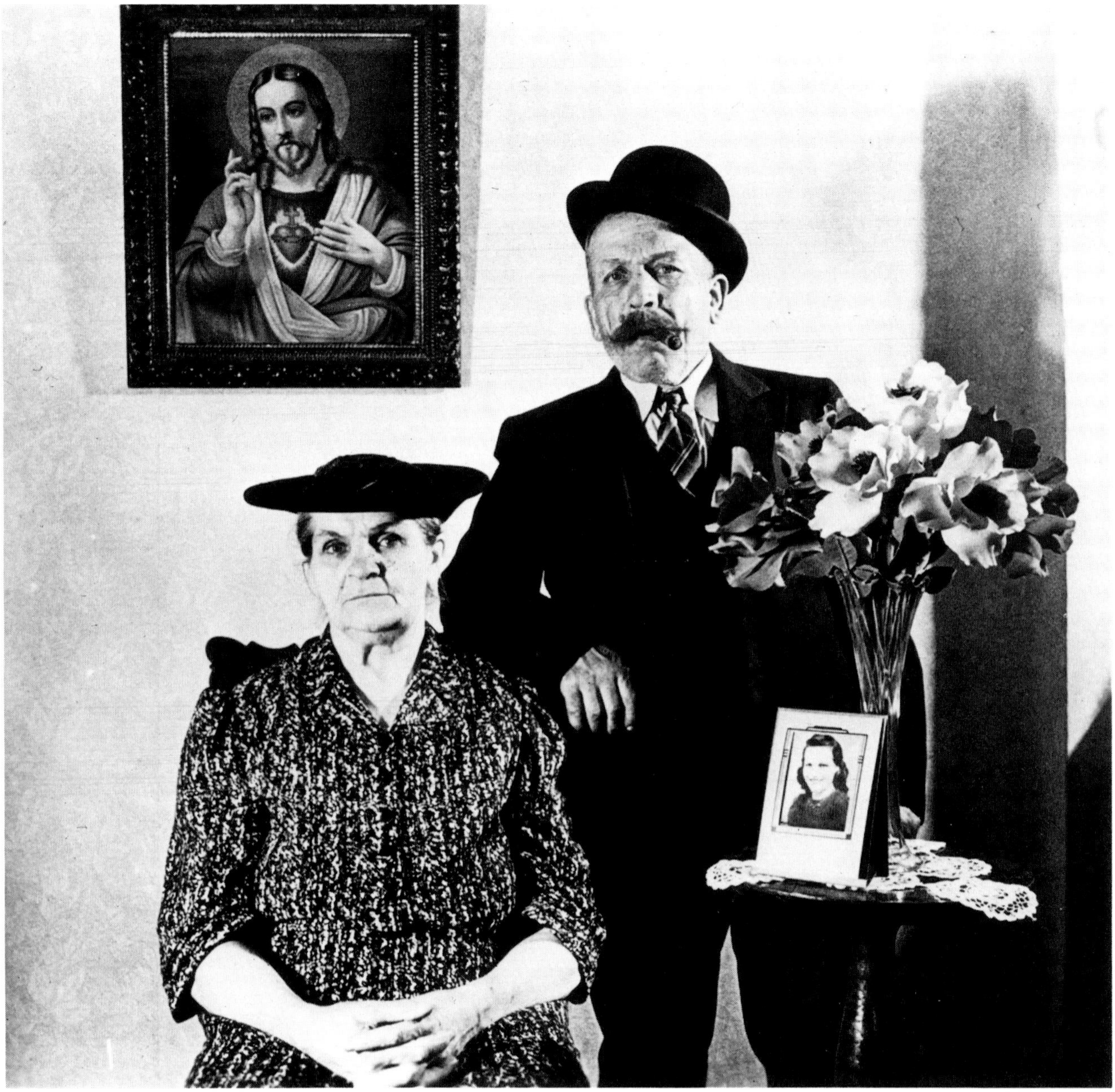

38. On Sundays the Scavnickys dress for church.